Topics in Mining, Metallurgy and Materials Engineering

Series editor

Carlos P. Bergmann, Porto Alegre, Brazil

"Topics in Mining, Metallurgy and Materials Engineering" welcomes manuscripts in these three main focus areas: Extractive Metallurgy/Mineral Technology; Manufacturing Processes, and Materials Science and Technology. Manuscripts should present scientific solutions for technological problems. The three focus areas have a vertically lined multidisciplinarity, starting from mineral assets, their extraction and processing, their transformation into materials useful for the society, and their interaction with the environment.

More information about this series at http://www.springer.com/series/11054

Ivan Kurunov · Aitber Bizhanov

Stiff Extrusion Briquetting
in Metallurgy

Ivan Kurunov (Deceased)
Novolipetsk Steel Company
Lipetsk
Russia

Aitber Bizhanov
Moscow
Russia

ISSN 2364-3293   ISSN 2364-3307 (electronic)
Topics in Mining, Metallurgy and Materials Engineering
ISBN 978-3-319-89200-9   ISBN 978-3-319-72712-7 (eBook)
https://doi.org/10.1007/978-3-319-72712-7

Printed on acid-free paper

This Springer imprint is published by Springer Nature
The registered company is Springer International Publishing AG
The registered company address is: Gewerbestrasse 11, 6330 Cham, Switzerland

The book is dedicated to the memory of my friend and teacher, Ivan Kurunov.

Preface

The fate of stiff vacuum extrusion (SVE), this "ugly duckling" of ferrous metallurgy is playing out in a surprising way. In a brief and clear way and yet without being noticed, it has made its debut in the Bethlehem Steel blast furnace (BF) in the United States of America and has made a triumphant return, after almost 20 years, to the lineup of industrial agglomeration technology. Despite the fact that it would not be entirely accurate to compare the scale of the blast furnace operation of what was once one of the largest American steel companies and that of the small, Indian company Suraj Products Ltd (Rourkela, West Bengali), it is this same experience that has enabled the authors to mold SVE into a fully fledged metallurgical technology. Six years of experience of using briquettes in the charge of a small blast furnace in Rourkela served as the basis of the technical and economic foundation of a project to construct a briquette factory at the Novolipetsk Steel Plant, which is designed to produce 700000 tons of briquettes annually (the deadline for commissioning this plant is 2018). Over the course of all this time, the briquette manufacturing plant in Rourkela (in West Bengal) has been a peculiar kind of Mecca for metallurgists who specialize in the field of the briquetting of natural and anthropogenic raw materials. Blast furnace specialists from the Novolipetsk Steel Plant, the West-Siberian Metallurgical Plant belonging to EVRAZ, the Metinvest Group (in the Ukraine) as well as delegations from Paul Wurth, and Tata Steel and many other Russian, and overseas specialists have familiarized themselves with the work of the plant. We would like to express our gratitude to the owner of the Suraj Products Ltd Company Mr. Yogesh Dalmia who is a real enthusiast of extrusion agglomeration technology and contributed personally to the development of the 100% briquetted operation of small-scale blast furnaces concept.

Within the confines of the project to construct a manufacturing plant at the Novolipetsk Steel Plant, the possibility of a partial replacement of sinter in the charge of a blast furnace with extruded briquettes (brex) is under examination. The practical realization of this substitution requires a growth in the production of brex of several orders of magnitude. The benefits of this "peaceful coexistence" of SVE and agglomeration are obvious. The efficiency of the agglomeration process itself increases as a result of the removal of fines from the charge for sintering. If 50% of the sinter were replaced in the charge of a BF with brex, the basicity of the remaining sinter should be between 2.8 and 3.2. It is worth noting that given this basicity, calcium ferrite should predominate in the structure of the sinter, providing the increase of its strength. The results of a mathematical model of BF smelting using iron ore concentrate and coal brex with a carbon content in the brex of 9.5% demonstrated a high level of efficiency of a partial (by 50%) replacement of the production of sinter by the production of brex with a reduction in the coke rate of 15% and a reduction by 50% of the dust emissions, CO_2 emissions at the sinter plant will decrease by 32% and sulfurous gas emissions by 43% during the manufacture of the sinter.

In 2017, several brex producing factories for ferroalloys industry were constructed in different countries of the world. The design of all these factories, for the most part was based on the results of that, which is described in this book, including the methods for the preparation of the charge for extrusion agglomeration. The first smelts already confirmed the justification behind our recommendation for the brex as an effective charge component for submerged electric arc furnaces. The share of brex in the charge of such furnace can exceed 50% of the charge, and the resulting alloy is distinguished by its higher quality. At present, the possibility of using brex based on chromium-containing materials in the charge of direct current electric furnaces is being studied.

It would be unlikely that there would have been the growth in the popularity of stiff extrusion briquetting around the world, or the construction of new plants for the manufacture of brex, were it not for the systematic improvement of the technological characteristics of stiff extrusion, or the search for new binding materials, and the optimization of the operating conditions, and servicing of extrusion equipment. The only producer of the stiff extrusion briquetting equipment in the world is the J.C. Steele&Sons, Inc. (North Carolina). We experienced beginner's luck in researching the metallurgical properties of the extruded agglomerates. We are undoubtedly grateful to Mr. Richard Steele (Vice-President), who kindly made laboratory equipment of the Company available to us. Specifically, this relates to a unique computerized laboratory extruder that was able to model all the significant stages of SVE. We are grateful to Jim Falter for the comprehensive reports on testing of our customers materials in the lab.

The authors also consider it their duty to express their gratitude to Tatyana Malysheva, a known Russian scientist and one of the authors of metallurgical mineralogy, with whose cooperation it was possible to build the theoretical models on the nature of the hot strength of brex as they are heated in a reducing atmosphere. The successful spread of SVE in the ferroalloys industry became possible thanks among other things to cooperation with the known Russian scientists and metallurgists, V.Ya. Dashevskyi (of the A. A. Baykov Institute of Metallurgy and Materials Science of the Russian Academy of Sciences) and A.V. Pavlov (of the National University of Science and Technology, MISiS), for which we also express our gratitude.

Moscow, Russia   Aitber Bizhanov

Contents

About the Authors

Ivan Kurunov was born on January 27, 1939, in settlement Satka, Chelyabinsk region, Russia. He graduated from Chelyabinsk Polytechnic Institute of Metals in 1961. Since 1963 till 2005, he is postgraduate student, Associate Professor, Professor in Moscow Institute of Steel and Alloys. Since 2005 up to August 2017, he is Chief Expert of Ironmaking division in Novolipetsk Steel Company (JSC NLMK).

He is author and co-author of more than 300 technical and scientific publications, including 8 books and brochures and more than 120 patents for inventions in metallurgy.

The very day it was decided that this book was to be published, my friend and teacher Ivan Kurunov was gravely injured in a traffic accident. He died on August 31st, the day that the press service of Russia's largest steel producer—the Novolipetsk Steel Company—announced the start of the construction of a new extrusion briquettes manufacturing factory with a capacity of 700,000 tons per annum. This project is Ivan's brainchild. Over the past six years, we have worked on this project together with our colleagues in Russia, the United States and India, side by side and, sometimes, back to back, in what we called "all-round defence".

The project is based on Ivan's ideas and the results of joint research conducted in Russia, the United States, and India. It is our deep conviction that stiff vacuum extrusion technology is the long-awaited key to quest to find an environmentally friendly and effective method to agglomerate natural and anthropogenic raw materials for ferrous metallurgy.

Ivan Kurunov's name will remain known in the history of metallurgy for his pioneering works in the field of blast-furnace production and efficient recycling of iron-containing materials.

Aitber Bizhanov was born on October 6, 1956, in Buynaksk, Russia. He graduated from the Moscow Physical-Technical Institute in 1979. He spent 11 years as a Senior Researcher with the Institute for High Temperatures, USSR Academy of Sciences. In 1992, he joined EVRAZ, a large, vertically integrated steel and vanadium enterprise with global assets. He joined Kosaya Gora Iron Works in 2005 as Commercial Director, introducing a briquetting technology for recycling of the natural and anthropogenic materials. Today, he is an Independent Representative of the J.C. Steele & Sons, Inc. company in Russia & CIS, Eastern Europe, and Turkey (since 2010). He has Ph.D. in the agglomeration of natural and anthropogenic materials in metallurgy. He is author of more than 50 publications in international journals, one book on briquetting (in Russian) and owner and co-author of 12 Russian patents in the field. He is author and owner of the "BREX" trademark.

Chapter 1
Introduction

1.1 A Brief History of Industrial Briquetting in Ferrous Metallurgy

The development of ferrous metallurgy, the growth of the capacity of blast furnaces (BF) and increased ore consumption required the creation of technology for the agglomeration of natural and fine anthropogenic materials generated during the course of the entire process from iron ore extraction through to iron smelting. The first industrial agglomeration technology, which appeared back in the nineteenth century, was briquetting technology—production of lumps of regular geometric shapes from iron ore fines by applying pressure using different binders or without them, with further drying and firing or curing under natural conditions. Industrial briquetting began with the first commercially successful production of briquettes from fine magnetite iron ore in 1899 in Finland [1]. Moisturized magnetite concentrate briquettes were produced without application of any binder and using clay brick making equipment with a pressure of 30–50 MPa. Energy consumption for the production of 1 ton of briquettes was 5 kWh. Raw briquettes (150 × 150 × 75 mm) were subjected to strengthening roasting in a tunnel kiln at temperatures of up to 1400 °C. In the process of roasting a sulfide sulfur oxidation occurred, which was further removed up to a level of 98% and the oxidation of magnetite to hematite took place. Despite the large size, these briquettes were successfully used in a low shaft blast furnace (BF) with a capacity of 50–140 t/day at a factory in Pitkäranta. Sintered briquettes (with a porosity of 40%) had a high mechanical strength (10 MPa) and a high reducibility and their use led to a reduction of Coke consumption and increased productivity of BF. The success of the project has contributed to the rapid spread of such technology. In 1913, there were already 38 similar briquetting lines in operation (16 in Sweden, 12 in England, and 6 in the United States).

Subsequently, for the production of briquettes lever, revolving, circular, conveyor, and roller presses were used. Depending on the type of material and binder

© Springer International Publishing AG 2018

I. Kurunov and A. Bizhanov, *Stiff Extrusion Briquetting in Metallurgy*,
Topics in Mining, Metallurgy and Materials Engineering,
https://doi.org/10.1007/978-3-319-72712-7_1

(if any) that were used briquettes were pressed with low (50–150 kg/cm^2), medium (150–750 kg/cm^2), or high (over 750 kg/cm^2) pressure [2]. The low productivity of the technological processes of briquetting as well as the insufficient strength properties of briquettes has prevented this technology from playing a meaningful role in the agglomeration of raw materials for blast furnace production. This role is now being played by the technologies of sinter production and pelletizing that were developed back in the early twentieth century. The first patents for the iron ore sintering method were obtained in Germany in 1902 and 1905, but the birth of a high-performance sintering process was in 1906 when A. Dwight and R. Lloyd in the United States patented a conveyor sintering machine [3]. The high technological effectiveness of the process and the possibility of recycling dispersed iron containing wastes, inevitably generated during iron and steel production, stimulated the development of this process and up to the present day the sinter strand sintering method is the main technological process for the agglomeration of raw materials in the steel industry, and is also widely used in nonferrous metallurgy. The main disadvantage of this agglomeration technology lies in its considerable adverse impact on the environment. A sinter plant within a metallurgical plant is the source of more than 50% of all dust and gas emissions.

The increase of lean iron ore and iron ore concentrate production from these has led to the development of the new technology for the agglomeration of fine concentrate. This technology was the production of pellets—pelletizing of iron ore concentrates with the formation of spherical granules 9–16 mm in size and their strengthening firing in a shaft furnace or on conveyor indurating machines. The patent for this technology was obtained by Andersen in Sweden in 1916, and the industrial production of pellets began to evolve rapidly from the second half of the twentieth century.

At the turn of the twentieth and twenty-first centuries in the steel industry, the technology of briquetting again began to be applied for the recycling of fine iron and iron-zinc-containing wastes—sludge, dusts, and fines. In Germany, technology has been developed for the production of briquettes from iron-zinc-containing sludge with their subsequent smelting in the cupola—Oxy-Cup, which is operated using an enriched oxygen blast [4]. Briquettes are being produced using vibropressing technology with Portland cement as a binder. The metallurgical quality of such briquettes meets the requirements of the BF charge. These briquettes qualify for the blast furnace in terms of their metallurgical properties. Vibropressing technology is applied in metallurgical plants in Sweden, where the BF are operating on 100% pellets and briquetting is being used as recycling technology for the agglomeration of flue dust and pellets fines, and carrying out recycling of the flue dust and pellet screenings. Briquetting by vibropressing began to be applied in the United States, Russia, and other countries, and not only for waste processing, but also for the agglomeration of fine ores and concentrates. The presence of carbonaceous materials (flue dust, blast furnace sludge) in the charge for briquetting ensures the self-reducing nature of briquettes, which reduces coke consumption during smelting proportionally to their amount in the charge of a blast furnace.

In addition to vibropressing technology roller presses are used for the production of briquettes from metallurgical anthropogenic materials. Use of organic binders ensures a high cold strength for the briquettes but restricts their application solely to steelmaking processes. The use of such briquettes in BF is possible only in very limited quantities due to the fact of their destruction when heated. To ensure high hot strength of roller-pressed briquettes a cement binder should be applied in an amount that is considerably higher than in the manufacture of briquettes using the vibropressing method.

At the beginning of the twenty-first century a new agglomeration technology of natural and anthropogenic raw materials appeared—Stiff Vacuum Extrusion (SVE), which for many decades was used for the production of bricks. The debut of the SVE in the preparation of agglomerated raw materials for blast furnaces took place in the 90s of the past century, but was not distributed widely. Auger extrusion for agglomeration of ore and metallurgical wastes first occurred in the 1990s, when Bethlehem Steel commissioned a stiff extrusion line for briquetting 20 tons per hour of sludge and flue dust [5]. These briquettes were used in Bethlehem Steel's blast furnaces. The line operated until 1996 when it ceased due to the closure of the plant. This milestone was not thoroughly investigated by metallurgists until 2010, when the authors of the present book began to study the characteristics of stiff extrusion and their influence on the metallurgical properties of extruded briquettes. By April 2011, more iron-making specialists were evaluating industrial briquetting plant production, along with the use of stiff extrusion briquettes as a major component of the blast furnace charge. These briquettes have become known as BREX (BREX-Extrusion Briquette) in technical literature [6].

1.2   Briquetting Technology and Metallurgical Properties of Briquettes

Briquetting is the creation of a solid structure of fine natural and anthropogenic materials using adhesives and/or heat strengthening treatment. Briquettes, as a component of the charge for metallurgical aggregates of ferrous metallurgy, must have certain metallurgical properties. The most important of these are cold strength, allowing them to preserve their integrity so they do not crumble with the formation of fines during transportation, discharging, and charging into the furnace; hot strength—the ability to maintain the integrity without disintegration in the process of heating and reduction; and reducibility.

A theoretical interest in briquetting technology followed by strengthening firing, as in the Gröndal project [1], remained until recently. Metallurgical properties of such briquettes (with a basicity of 1.4), produced on an industrial briquetting roller-pressing line, were studied in comparison with the properties of fired pellets with the same basicity [7]. The briquettes firing took place in an electric furnace in an oxidizing atmosphere at temperatures ranging from 1190 to 1250 °C.

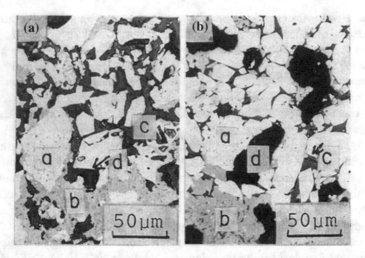

Fig. 1.1 Microstructure of briquette "**a**", fired at 1250 °C and of pellet—"**b**" (a—hematite, b—calcium ferrite, c—slag phase, d—pore)

Petrographic analysis revealed the similarity of the microstructure of briquettes and pellets (Fig. 1.1).

However, such technology was not widespread in metallurgy. The only methods for the thermal processing of briquettes, currently applicable on an industrial scale, are drying and heat-moisture treatment of vibropressed briquettes. Recently a fundamentally new concept of so-called hot briquetting has been developed. A mixture of fine iron ore and coal is heated to temperatures of 350–600 °C and pressed after the carbon reaches plasticity [8]. It is clear, however, that this reduces the economic efficiency of the technology in general as a result of the expenditure of energy for heating, the absence of which is an important competitive advantage of cold briquetting.

1.3   Roller-Press Briquetting

In the 20s of the past century roller presses were widely used for the briquetting of iron ore fines and metallurgical wastes. The frame of these presses has one or two pairs of rollers with steel sleeves on them. These sleeves consist of segments with forming cells. Pressing of the charge occurs in the gap between the rollers as they rotate towards each other. The capacity of modern roller presses can exceed 50 tons per hour and the pressure applied—up to 150 MPa.

The structure of the material pressed by roller presses undergoes changes associated with an elastic and irreversible deformation, with the destruction of the particles of the pressed material and with the formation of cracks within it. The release of energy of the elastic deformation after removal of the applied load can lead to the growth of the volume of the briquettes. The short duration of the mixture compression process can lead to "pressing" of the air in the briquette's body, which also reduces the strength of a briquette. In the process of roller-pressed briquetting the homogenization of the mix does not take place and the structure of briquette follows the structure of the mixture in the forming cell.

Roller-press briquetting is accompanied by the generation of a sufficient amount (up to 30%) of wastes in the form of a charge mixture, which does not enter the forming cells.

A specific feature of roller-press briquetting are the restrictions imposed on the moisture content of the briquetted material (less than 5–10%) which means that the charge should be subjected to dewatering and/or drying and corresponding facilities should be included in the briquetting plants. The required low moisture content of the charge restricts the utilization of hydrating binders, such as cement. The disadvantages of this briquetting technology are also related to the incomplete closing of forming cells, leading to a deterioration of the strength of the briquette, and with the lack of exposure of the briquette body under pressure, which degrades its properties and leads to the need for so-called "prepressing". However, roller-pressed briquettes were widely used in the early twentieth century in the charges of small blast furnaces and low shaft blast furnaces. Their share in the charge in the BF in the West Kalbe plant reached 30–40%, and in the low shaft BF in the Max Hütte Plant (Germany)—100%. These briquettes were made from iron ore, limestone, and coke breeze [9]. At that time roller-pressed briquettes were also used in the BF of Kushvinskyi's plant (up to 25% in the charge), as well as in the charge in the BF in the Kerch (100 thousand tons per year) and Taganrog metallurgical plants [10].

Interest in roller-press briquetting of raw materials for BF was renewed in the 60–70s in the past century due to the growing need for and relevance of the recycling of accumulated iron containing wastes. The Briquetting factory "Donetskaya" was built in 1961 in Donetsk (Ukraine). Similar briquetting lines at Alchevsk iron works (Ukraine), Nizhny Tagil and Magnitogorsk Iron and Steel works (Russia) were also commissioned. Experience of the operation of briquetting plants owned by Harsco Metals (United States) [11], producing briquettes for blast furnaces in the United States, Europe, and Asia (the total production volume is 1.5 million tons per year), has shown that the achievement of the required metallurgical properties of briquettes for BF became possible only when using a complex multicomponent binder (composition is not disclosed), whose share in the mass of the briquette reached 15–18%. This explains, in particular, the low content of briquettes in the blast furnace charge (10–15 kg per ton of iron). The same low percentage of roller-pressed briquettes is supported in the BF charge used by the ILVA Company (in Italy). Briquettes are produced by Primetals (Austria) with the use of lime and molasses as a binder [12].

The application of roller briquettes in blast furnace production did not enjoy widespread application, not only because of the high mineral binder content, which reduced the iron content in them, but also because of the substantially lower capacities of roller presses (up to 1200 tons per day) compared to the performance of the sintering machines (1500–15,000 tons per day) and conveyor indurating machines (2500–19,000 tons per day). However, even in the Ferroalloy industry, where the output and quality requirements of briquettes are much lower than in blast furnace production, roller-pressed briquettes have not been widely used, despite the positive results of full-scale testing of briquette utilization in the charge of ore-smelting furnaces for Silicomanganese production at the Zestafoni Ferroalloys plant, held in the 60s of the past century [13, 14]. Roller-pressed briquettes for industrial experiments were produced by a roller press with a capacity of 5 tons per hour from the Manganese ore with a particle size ranging from 0 to 5 mm using sulfite-alcohol vinasse as a binder (8% of mass) with a density of 1.2 g/cm^3. The testing furnace operated normally and steadily during testing with decreased power and a reducing agent consumption. The results of testing have shown that manganese ore briquettes are of adequate strength and are suitable for use in the charge of Ferroalloy furnaces.

In 1970, scientists at the Dnipropetrovsk Metallurgical Institute (Ukraine) examined the results of silicomanganese production with roller-pressed briquettes and with sinter in the charge of ore-smelting furnaces. The briquettes were made of a mixture of oxide manganese concentrates of I and II grades of Nikopol ore deposit (Ukraine) in equal amounts. The same mixture of raw materials was used for sinter production [15]. Prior to briquetting the charge was pulverized (from 0–10 mm to 0–3 mm). Briquettes were produced at a semi-industrial roller-press at a pressure of 500 kg/cm^2. A mixture of bitumen, mazut, and sulfite-alcohol vinasse was applied as a binder (10% of the weight of the briquetted mixture). The batch was mixed in steam-heated containers. Two types of briquettes were manufactured and smelted: with an excess amount of reducing agent (a mixture of concentrates—54.5%, fluvial sand—9.1%, coal—27.3%, bitumen and mazut mixture—3.6%, and sulfite-alcohol vinasse—5.5%) and with the stoichiometric amount of reducing agent required for the reduction of silicon and manganese (a mixture of concentrates—60.6%, fluvial sand—10.1%, coal—20.2% coal, bitumen and mazut—3.6%, and sulfite-alcohol vinasse—5.5%). Laboratory tests have shown the benefits in terms of the strength properties and thermal stability of raw briquettes over the fired ones. It was decided to produce the silicomanganese with raw briquettes in the furnace charge. Experimental smelting with briquettes with the composition indicated above and melting with sinter were carried out in a three-phase open ore-smelting furnace with a capacity of 1.2 MVA at a voltage of 81.60–85.89 V and a current of 6000–6500 A in a continuous process with a closed furnace top. Briquettes with an excess of reducing agent showed enhanced electrical conductivity and when smelted gave a significant increase in current load, which led to the rise of the electrodes and to the opening of the furnace top. Smelting of silicomanganese with briquettes containing a stoichiometric amount of reducing agent in the charge was achieved without any complications. The load was kept steady; nothing else was fed into the furnace but the briquettes. The results of this comparative study showed that smelting of silicomanganese on briquettes is more efficient than smelting with sinter.

In the article [16] the results of the development and commercialization of technologies for the briquetting of a mixture of manganese ore concentrates and bag-house dusts and of the subsequent use of these briquettes for ferroalloys production are set out. The need to recycle these dusts stems from the high volume of their generation (up to 50 thousand tons annually). It has been established that to obtain mechanically strong briquettes (with a specific compressive strength of 8–12 MPa) the optimal parameters of briquetting should be as follows: a moisture content in the blend ranging from 4 to 6%, a binder content (sulfite-alcohol vinasse) of 6–8%, while the share of dust and sludge should be at 30%, with a minimum compressive pressure of 19.6 MPa. The kinetics of the reduction of ore-dust briquettes and of manganese sinter was studied comparatively. It was found that the briquettes in the charge provided the maximum degree of reduction at different temperatures. The use of silicon and manganese from waste materials was increased due to the presence of bound carbon within them, which prevented the formation of silicates inevitably generated in sinter. Results of the smelting of high-carbon ferromanganese melts showed that the productivity of the furnace in which briquettes were used increased from 73.33 up to 75.67 t/day, and specific electricity consumption decreased by 90 kWh/t. Consumption of coke breeze decreased by 34 kg/t. The presence of a carbonaceous reducing agent in briquettes dramatically increases the temperature of the softening of the briquettes up to 1250–1400 °C. Manganese ore fines briquettes manufactured using sulfite-alcohol vinasse as a binder had a softening temperature ranging from 750 to 850 °C.

1.4 Vibropressing Briquetting

In the 70s of the past century vibropressing technology began to be used for the agglomeration of natural and anthropogenic raw materials for ferrous metallurgy. The briquettes according to this method are pressed at low pressure (0.02–01 MPa), and with a simultaneous vibration (a frequency of 30–70 Hz and an amplitude of 0.2–0.6 mm) of the pressed mixture. Multicell molds are used for briquetting with dimensions ranging from 20 × 20 × 20 mm up to 500 × 1500 × 1500 mm. The pressing cycle lasts no more than 30 s. Vibration with a frequency exceeding 50 Hz, sharply reduces the viscosity of the mixture and internal friction between particles, which facilitates their convergence. During the vibration air and excess moisture are forced out of the mixture. All this together leads to a greater degree of compaction of the briquettes at lower pressures than during compression [17]. Portland cement is widely used as a binder in vibropressing briquetting. Its share can reach 10–15% [18].

The capacity of vibropressing equipment depends on the size of the briquettes that are produced, and does not exceed 30 tons of raw briquettes per hour. Green vibropressed briquettes have a very low mechanical strength, which does not allow them to be transported by conveyor to the warehouse, and requires special arrangements for the stacking of pallets with these briquettes on the shelf and their subsequent transport to the heat-moisture treatment chamber. One of the first commercial vibropressing briquetting lines was put into operation at the enterprise of SSAB Company in Oxelösund (Sweden) [19]. Briquettes that are 60 × 60 mm in size and hexagonal in cross section are used in the blast furnace (60–100 kg of briquettes per ton of iron). Composition of the briquettes, expressed as a %: 25—flue dust; 50—a mixture of scrap, BOF sludge, pellets fines; 5–8—aspiration dust, 10–12—Portland, 5–8—moisture. The particle size of the components of a mixture ranges from 0 to 5 mm. In March 2012 in Finland at the Raahe Works Company, a vibropressing briquetting line for the production of briquettes for blast furnaces was commissioned after the closing of the sinter plant [20]. Briquetted charge consisted of: mixture of pellets fines (71,000 tons), coke dust (12,000 tons), premix of flue dust and steel scrap (59,000 tons), mill scale (59,000 tons), briquettes fines (34,000 tons) and scrap (15,000 tons), and slag cement (17,000 tons) and rapid cement (19,000 tons). The total amount of briquettes recycled via blast furnace process is 283,000 tons per year. Briquette size is 60 × 60 mm and weight is 475 g. After the 48 h of strengthening the cold strength of briquettes according to ISO 4696 was 74% (the share of pieces larger than 6.3 mm). Coke rate was decreased by 6% by feeding 120–130 kg of briquettes per 1 ton of hot metal into the charge of the BF due to the fact that the briquettes contained carbon. It is seen that vibropressing led to the formation of significant quantities of briquettes fines (34,000 tons per year). It was also impossible to agglomerate efficiently BOF sludge (60,000 tons per year; moisture content 33–39%) and BF sludge (30,000 tons per year; moisture content 20–25%) by vibropressing. In 2017, these materials were tested by SVE agglomeration method. It was concluded that the brex made of BOF sludge, pellets fines, and coke dust might be suitable for the blast furnace.

At another SSAB enterprise vibropressed briquettes are manufactured from the mixture of pellets fines, dusts, and sludge from BF and from steelmaking units. Briquettes with two different compositions (93.5% of pellets fines, 6.5% of Portland cement and 62.7% of pellets fines, 30.8% of sludge and dusts, and 6.5% of Portland cement) are produced using the vibropressing method. One peculiarity of these briquettes is a tendency to swell considerably during the reduction process using carbon monoxide at 950 °C, which is linked with a similar abnormal swelling of the indurated pellets [21–23]. The mechanism of the catastrophic swelling of iron ore pellets is associated with the growth of iron whiskers during wustite reduction, which is enhanced in the presence of CaO [24, 25]. The degree of swelling decreased when adding hydrogen to the reducing gas. Something that was also noted was the fact that the tendency to swell in the presence of cement is reduced when its content in the body of a briquette exceeds 10% (mass). The largest degree of briquette swelling took place with a cement share in the mass of the briquette ranging from 4 to 6%. It was noted that at high temperatures the converted phases of cement, olivine, and wustite react with the formation of $(Ca, Mg, Fe)_2SiO_4$ (dicalcium silicate, forsterite, and fayalite). The properties of this phase are determined by the relationship between the values content of pellets fines and cement. The formation of the silicate melt at a lower temperature is also possible (1150 °C). These briquettes were also tested in the experimental blast furnace. It is claimed that unlike testing of the briquettes in laboratory conditions, their reduction in the experimental blast furnace was not accompanied by catastrophic swelling. Reduced briquettes, after extraction from the experimental blast furnace, had a significant wustite core even with fairly profound lowering in the furnace. For the explanation, authors indicated the following possible reasons: a large volume of briquettes, creation of a slag system $CaO-SiO_2-MgO-Al_2O_3-FeO$, and coating of the wustite grains in a molten slag.

Application of the Blast furnace simulator for the study of the behavior of vibropressed briquettes was described in article [26]. The main components of briquettes are—mill scale (41.2%), various scraps (23.4%), flue dust (9.5%), aspiration dust from the dosing and casting units (4.8%), coke dust (6.4%), pellets fines (2.8%), and other anthropogenic materials. Three different combinations of binders were applied: 10% Portland cement; 8% Portland cement +3% crushed granulated slag; 8% of Portland cement +6% crushed granulated slag. Briquettes were manufactured on an industrial vibropressing line, however, for blast furnace simulator tests samples of briquettes were cut into bars with a cross section in the shape of a triangle (Fig. 1.2).

It is clear, however, that in this case, it was impossible to obtain authentic test results, because the specific surface of the briquette sample (the ratio of surface area to volume) was higher than in the original briquette and hence the speed of reduction was different. The following results were obtained in the tests:

(1) Briquettes showed a significantly faster reduction compared with pellets in the temperature range of 780–1100 °C, which can be explained by the presence of carbon in their composition and by a higher porosity, which, according to the authors, was due to the effects of hematite–magnetite phase transition;

Fig. 1.2 The location of fragments of briquettes and pellets in a basket for a blast furnace simulator [26]

(2) Swelling of briquettes (25–50%) was observed during the wustite–iron phase transition at 900–1000 °C;
(3) The addition of granulated slag had a slight increasing effect on the swelling of the briquettes. The effect was more pronounced with a 3% addition than with a 6% addition;
(4) The structure of the briquette does not disintegrate in BF conditions even though the cement phases decompose in the high temperature of the BF process. The phase transformation of $Ca(OH)_2$ first to $CaCO_3$ and secondly to $Ca_2Fe_2O_5$ may have an effect on this behavior.

In Russia, the first vibropressing briquetting factory was commissioned at the JSC Tulachermet in 2003 [27]. The capacity of the line enabled a monthly production of up to 8000 tons of briquettes. Two types of briquettes were used as charge components of a BF: so-called "washing" briquettes that contained iron and carbon. It is known that in some cases, technologists were faced with "cluttering" of the hearth of the blast furnace due to the deterioration of coke's ability to filter when filling the voids between its particles with pieces of slowly moving smelting products. This can result in combustion of air tuyeres, decrease of hearth heating, and other phenomena, which reduce the melting performance. One of the most effective means of combating this phenomenon is "flushing" the hearth using liquid slag containing FeO or MnO. Typically, the special sinter made using mill scale serves as the "washing" material. Mill scale briquettes can also be used for such purposes. The value for the design compressive strength was supposed to be at least 6.0 MPa. After drying its values were—3.83 MPa, after heat-moisture treatment—6.9 MPa.

In total, during the course of the existence of the line more than 52 thousand tons of briquettes were manufactured (about 50 thousand tons of briquettes containing iron and carbon, and 2700 tons of washing briquettes). During the unloading of the briquettes from the bunker their bridging was noted [28]. The maximum shares of the briquettes in the charge of the blast furnaces were as follows: BF No. 1–32 kg/t of hot metal, BF No. 2–56 kg/t of hot metal. With the application of briquettes flue dust blowout decreased by 4 kg/t.

Factor analysis showed that with the application of iron and carbon containing briquettes the specific consumption of dry skip coke decreased by 14.4 kg/t of hot metal, which corresponded to the coke's replacement in the body of the briquettes with coke breeze by a ratio of 1 to 1. Pig iron production increased by 37 tons/day, which is associated with an improvement in the gas permeability of the charge column due to the improved granular composition of the charge. This helped to increase the blast volume and, as a result, to intensify melting.

After several years of operation, the briquetting factory was closed as a result of changes in economic conditions affecting the availability of suitable carbonaceous materials for briquetting.

A series of campaigns on the use of vibropressed briquettes of varying compositions on a cement binder in the charge of blast furnace with a volume 1000 m^3 was carried out in 2003–2004 by the Novolipetsk Steel Company (JSC NLMK) [29–31].

Table 1.1 Chemical composition of sludge and coke briquettes

Fe$_{total}$	FeO	Fe$_2$O$_3$	CaO	SiO$_2$	Al$_2$O$_3$	MgO	C	ZnO
35.1	30.43	16.0	19.9	7.1	1.36	1.12	19.0	1.5

In the first stage for the evaluation of the effectiveness of the iron-zinc-containing sludge recycling 2500 tons of briquettes were smelted. The composition of the briquettes was as follows: 65% converter sludge, 20% coke breeze, and 15% Portland cement binder. The laboratory tests for optimization of the composition of the briquettes conducted later proved that their carbon content (Table 1.1), was oversized [32]. The coke rate in the (10 day) period of trial melts varied over a range from 50 to 70 kg/t during the first 5 days up to 190 kg/t during the final few days and 121 kg/t of hot metal was produced on average. The economic performance of the furnace operation in trial and in reference periods and the results of the computer simulation of the BF melt using sludge and coke briquettes are shown in Table 1.2.

Table 1.2 BF melt performance for furnace operation with or without sludge and coke briquettes

BF furnace performance	Reference period 1–15.03.2003 31.03–6.04.2003	Trial period 20–30.03.2003	Simulation results[a]
Productivity, t/day	2050	1828	1851
Dry skip coke rate, kg/t	482.5	485.5	459
Natural gas consumption, m^3/t	86.1	80.2	82
Briquettes rate, kg/t	–	121	125
Fe content in the burden, %	59.1	58.06	58.4
Blast temperature, °C	994	966	970
O$_2$ content in the blast, %	26.8	26.5	26.5
Top gas pressure, MPa	0.101	0.098	0.101
Si content in the hot metal, %	0.72	0.84	0.75
Reduced[b] productivivty, t/day	1931	1828	1819
Reduced[b] coke rate, kg/t	508.7	485.5	470

[a]Computer simulation of BF melt using a mathematical model of the BF process; [b]Reduction to comparable conditions is carried out using a factor-by-factor analysis method. All in all, the results of trial melts demonstrated the practical possibility of an efficient implementation of the new type of lumpy fluxed iron-bearing material, which, due to the carbon content within it is self-reducing, and its implementation results in a pro rata decrease in the coke rate

Table 1.3 Chemical composition of industrial briquettes made from iron ore concentrate

Fe	FeO	Fe_2O_3	CaO	MgO	SiO_2	Al_2O_3	C
42.3	16.4	42.1	9.8	0.8	6.0	1.1	17.2

In the course of the smelting of the briquettes the coke to coke breeze replacement factor was 0.96 kg/kg, and the replacement factor for the carbon in the coke to the carbon in the briquettes was 0.88 kg/kg. The decrease in furnace productivity during the smelting of the briquettes, apart from the decrease of the iron content in the burden, was induced by an increased physical and chemical heating of the melt products, due to an inadequate adjustment of the burden-weight ratio together with the increased rate of the carbon-bearing component in the burden, which was not well known to BF process engineers. The furnace productivity was also influenced by a high basicity, together with a high-melting slag from the waste ore in the briquettes, which was slowly assimilated by ferrous primary slag formed of sinter and pellets, which as a result increased the slag viscosity.

In the second stage, 2475 tons of briquettes made of a mixture of iron ore concentrate, coke breeze, and Portland cement were smelted. The aim of the trial melts using briquettes made of iron ore concentrate was to estimate the feasibility and efficiency of the smelting of cement-bonded briquettes in large quantities for the maximum replacement of coke with coke breeze, which serves as a reductant in the briquettes. Magnetite iron ore concentrate was used for briquette production. These briquettes (Table 1.3), which are the same as the briquettes made from BOF sludge, were produced with an oversized fraction of coke breeze (20%) and cement (15%).

Smelting of a trial lot of iron ore and coke briquettes between August– September 2003 was carried out in three stages with a gradual increase of the briquettes rate (122, 198, and 303 kg/t of hot metal). Fractional analysis of the BF burden in the course of the briquettes smelting essentially differed from the fractional analysis of the burden during the reference period and in essence varied with the briquettes rate increase. This is primarily relevant for pellets, for which the fraction in the burden during the reference period was 23%, and during the briquettes smelting it decreased down to 12.7, 11.8, and 4.3%. The sinter fraction in the burden also fluctuated noticeably: 76.9, 80, 76.3, and 78.4%, accordingly (Table 1.4).

With respect to the overall impact of the briquettes rate on furnace productivity, apart from the decrease of Fe content in the burden, it is induced, in the same way as with sludge and coke briquettes, by the formation of a high-melting slag with high basicity from the waste ore of the briquettes, leading to a viscosity increase in all the primary slag.

The efficiency of the application of coke breeze, contained in the briquettes, was decreasing pro rata to their amount, which was related to the oversized content of carbon in the briquettes, which by 5% exceeded the stoichiometric amount (12.2%) required for reduction of iron oxide in the briquettes. The coke breeze that was not consumed for the reduction of iron in the briquette itself and iron in the primary slag, remained in the coke bed, decreasing its drainage capacity.

Table 1.4 BF performance during the smelting of briquettes of iron ore and coke breeze

Melting characteristics/furnace operation periods	21–25.08 6–10.09	26-30.08	31.08-02.09	02-04.09	26.08-04.09
Productivity, t/day	1908	1732	1781	1725	1743
Rate:					
Sinter, kg/t	1194	1325	1283	1348	1319
Pellets, kg/t	371	210	199	67	168
Briqettes, kg/t	–	122	198	303	192
BOF slag, kg/t	75	–	–	–	–
Fe content in the burden, %	58.83	58.02	57.3	55.98	57.26
Coke rate, kg/t	505	489	473	497	487
Natural gas consumption, m^3/t	74	70	76	79	74
Temperature of the periphery, °C	516	484	483	527	495
Bottom pressure drop, atm	0.869	0.860	0.880	0.866	0.867
Top pressure drop, atm	0.131	0.140	0.120	0.124	0.13
Content: [Si] in hot metal, %	0.74	0.85	0.64	0.86	0.79
Reduced productivity, t/day	1908	1780	1812	1828	1815
Reduced coke rate, kg/t	505	480	474	484	480
Replacement factor (coke to coke breeze), kg/kg	–	1.02	0.78	0.35	0.66
Amount of coke breeze in the briquettes, remaining in the furnace[a]	–	$\frac{20.0}{100}$	$\frac{33.5}{200.6}$	$\frac{49.3}{348.8}$	$\frac{31.7}{348.8}$

[a]Estimated amount of the excessive coke breeze in the briquettes (that is to say remaining in the briquettes after a complete reduction of ferrous oxide): in the numerator—t/day in the denominator—total amount of excessive coke breeze in the briquettes at the end of the period

Drainage capacity of the hearth was estimated with the help of the «DMI» calculation index [33]. The index should be calculated according to the following formula:

$$DMI = 2 \cdot T_{hm} - 120.62 \cdot [Si] - 128.40 \cdot [P] - 155.64 \cdot [S]$$
$$+ 10.89 \cdot [Mn] - 389.11 \cdot [C] - 190 \cdot B_{sl} - 690.46$$

The "DMI" calculation confirmed the accumulation of coke breeze in the coke bed. In the BF operating period after the smelting of the briquettes (6–10.09.2003) "DMI" decreased by 1.34 times against the operating period prior to the smelting of the briquettes (21–25.08.2003), showing values of 127 and 170 for these periods, respectively. The coke to coke breeze replacement factor was decreasing with the increase of the briquettes rate due to the reasons given above.

Industrial experience of the utilization of these briquettes in the BF melt enables a conclusion to be made that such briquettes, provided the optimum content of the carbon and the minimum cement content are just enough to reach the required strength of the briquettes, constitutes a full-fledged, ready self-reducing component of the BF burden, and its implementation ensures a decrease in the coke rate in the BF melt.

In the third stage, 2560 tons of vibropressed briquettes made of a mixture of BF sludge (59%), mill scale (20%), coke breeze (10%), and cement (11%) were smelted in a blast furnace with a volume 2000 m^3. Increased bridging has been observed during the discharging of the briquettes from the bunkers. In order to estimate the efficiency of such a recycling scheme, a trial lot of such briquettes was produced, which apart from BF sludge (59%) and cement (11%) contained rolling mill scale (20%) and coke breeze (10%). The average chemical composition of the briquettes is shown in Table 1.5.

These briquettes (2560 t) were smelted within a period of 11 days (from 29.11.2004 until 9.12.2004) in a BF with a useful volume of 2000 m^3. The average briquettes rate for the period was 62 kg/t with a daily variation ranging from 36 kg/t up to 81 kg/t. The efficiency of the smelting of the sludge and coke briquettes was evaluated by way of a comparison of the melting characteristics in the trial period and in the reference period of the operation of the BF. The average characteristics for the furnace operation periods before (6–10.11.04, 26–28.11.04) and after the smelting of the briquettes (11–15.12.04) were taken as the reference period characteristics (Table 1.6).

A slight reduction of BF productivity during the smelting of the briquettes was mainly induced by the decrease in iron content in the burden, as well as by the negative influence of excessive basicity and viscosity of slag, formed out of the waste ore of the briquettes.

With regard to the lower carbon in the coke to the carbon in the briquettes replacement factor, when compared to the value obtained during smelting of the BOF sludge briquettes (0.88 kg/kg), this can be explained by the fact that the carbon content in the briquettes significantly exceeded the stoichiometrically required value (9.3%).

Table 1.5 Chemical composition of the briquettes

Components	Fe$_{total}$	FeO	Fe$_2$O$_3$	SiO$_2$	CaO	Al$_2$O$_3$	MgO	ZnO	LOI
Average value	33.9	17.81	27.25	8.78	14.90	1.06	1.71	0.26	26.39

Table 1.6 Blast furnace operation parameters while smelting iron- and carbon-bearing briquettes made of BF sludge and mill scale

Furnace operating factors	Units	Reference period	Trial period
Furnace productivity	t/day	4097	3960
Briquettes rate	kg/t	–	62
Dry skip coke rate	kg/t	459	458
Oxygen content in the blast	%	26.6	26.5
Blast temperature	°C	1153	1154
Top pressure	atm.	1.45	1.39
Total level of gas utilization	%	44.9	44.3
Peripheral gases temperature	°C	359	417
Top pressure drop	atm.	0.175	0.170
Bottom pressure drop	atm.	1.128	1.107
Iron content in the burden	%	59.23	58.28
[Si] content in hot metal	%	0.76	0.69
Reduced[*] furnace productivity	t/day	4097	4056
Reduced[*] coke rate	kg/t	459	447.6
Coke to coke breeze replacement factor	kg/kg	–	
Carbon in the coke to carbon in the briquettes replacement factor	kg/kg	–	0.63

[*]Compared with the values of the basic period of BF operation before and after experimental smelting

The amount of carbon that is accumulated by the briquettes together with the carbon-bearing BF sludge would be quite sufficient for a complete reduction of contained iron [32].

The results of experimental melts confirmed the effective application of iron and carbon containing briquettes as a new blast furnace charge component. Due to the presence of carbon such briquettes are self-reducing and their utilization leads to a proportional reduction of coke consumption. The main results of the campaigns are as follows: vibropressing can provide the required values for the mechanical strength of briquettes made of natural and anthropogenic oxide materials with a share of the cement binder of not less than 8–10% by weight of a briquette; the value of the compressive strength for most of the briquettes was not less than 30 kgf/cm² and ensured their integrity during overloads and transportation with a less than 5–7% fines generation (−10 mm).

The results of the investigation of the metallurgical properties of the laboratory briquettes made by vibropressing, as well as of briquettes used in full-scale testing are described in [32]. In particular, the behavior of carbon-free briquettes made of magnetite concentrate (91.2%), Portland cement (8.8%), BOF sludge (91.9%), and Portland cement (9.1%) whilst undergoing heating in the reducing atmosphere has been studied. Since the cement binder maintains the strength of the briquette up to temperatures of 750–900 °C, the formation of new solid briquette microstructures was attributed to physical-chemical processes that occur in the body when heating the briquettes in a reducing atmosphere. Such processes are primarily a reduction of iron oxides in the solid-state reaction between wustite, oxides of cement stone, and gangue components of a briquette with the formation of calcium-iron-magnesium silicates. In the briquettes made of magnetite concentrate and converter sludge, a zonal microstructure with a surface layer of metallic iron has been optically determined. This fact is due to the relatively high density of briquettes, their considerable size and the lack of a solid reductant in their composition. In the briquettes made of iron ore magnetite concentrate a magnetite to wustite reduction process is clearly diagnosed throughout the volume of the sample, on the surface there is the layer of metallic iron with a thickness of 3–5 mm, forming a superficial shell (skeleton) of a briquette (Fig. 1.3, left). In the deeper layers (up to a distance of 20–25 mm from the surface of the briquettes) only small zones can be observed with particles of metallic iron on the edges of the wustite grains. In the central part of briquette, the metallic iron is missing and the entire iron containing phase is represented only by wustite and ferrous olivine. Depending on temperature and duration the conditions of a briquette in a reducing atmosphere in different parts of the briquette iron–silicate phase was either in a plastic or a near-plastic state and was filling the space between grains of ore in the form of a glass phase (Fig. 1.3, right). Due to the large content of SiO_2 in the concentrate (6.3%), the presence of impurities Al_2O_3, MgO in the mineral phases of cement, as well as the extended surface of contact between the concentrate (particle size 70–120 µm) and cement stone particles, a large portion of the active wustite—a product of magnetite reduction—reacts with silica and forms a system Fe_2SiO_4–Ca_2SiO_4 comprising an eutectic with a crystallization temperature of 1120 °C (phase diagram CaO–FeO–SiO_2 on Fig. 1.4). Iron–silicate phase has a low reducibility, which together with the diffusive difficulties and large size of a briquette explains the lack of metal iron in its central part. Conservation of the form of a briquette made of magnetite concentrate as it undergoes heating in a reducing atmosphere is provided by the formation of a surface metal iron frame with a thickness of 3–5 mm, which ensures

Fig. 1.3 Microstructure of a reduced briquette made of magnetite concentrate. Left: 1—metallic grains, 2—wustite, 3—olivine phase; right: 1—wustite, 2—olivine phase. Reflected light, magnification × 500

Fig. 1.4 Phase diagram of
CaO–FeO–SiO$_2$

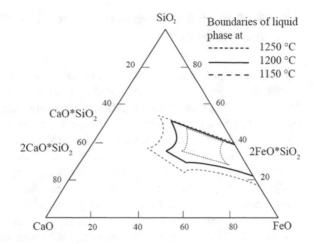

the strength of a briquette at the stage of dehydration of calcium hydrosilicates in the cement stone. According to the authors, the destruction of a briquette as it is heated up to 1150 °C is also prevented by the formation of an iron–silicate matrix of iron–calcium olivine in the full body of the briquette.

It is known that the system CaO–FeO–SiO$_2$ has a eutectic (25% CaO · SiO$_2$ and 75% FeO · SiO$_2$) with a crystallization temperature of 1030 °C. There are also fields with a crystallization temperature close to 1150 °C.

Apparently, because of the limited time in the furnace at a temperature of 1150 °C and above, the iron–silicate phase does not reach a liquidus temperature and remained in a plastic condition without violating the integrity of the briquette in the already established frame of metallic iron with a melting point above 1500 °C.

In 2010, the Kosaya Gora Iron works vibropressing briquetting factory (located in Russia) was commissioned for the production of 120,000 tons of briquettes from a mixture of iron ore concentrates and fines, and flue dust with a Portland cement binder (not less than 10% by weight of the briquette). The exposure time of the briquettes in a curing chamber (for heat-moisture treatment)—36 h. The density of briquettes ranged from 2.0 to 5.0 g/cm^3. Their compressive strength: not less than 3.5 MPa, and the moisture content was up to 9%. The proportion of briquettes in the ore part of the blast furnace charge—100 kg per ton of hot metal. Bridging was noted during loading [34]. In total, the factory has produced over 300 thousand tons of briquettes. There is almost no information available in open source material concerning a systematic analysis of the results of the operation of this factory. The factory has practically been out of operation since October 2015.

The results of full-scale testing and of the practical implementation of vibro-pressing briquetting technology confirm the possibility of achieving the required level in terms of the metallurgical properties of the agglomerated products. However, this technology creates some significant technological limitations, which are difficult to overcome, or leads to a significant increase in the cost of a briquette.

1.5   Stiff Vacuum Extrusion Briquetting. What's the Difference?

The first attempt to use extrusion for agglomeration of iron ore fines took place in the 50s of the past century. A hydrated mixture of iron ore fines and bentonite was filed in a chamber and then extruded in a vacuum through the die with the production of cylindrical briquettes [35]. However, this method had not enjoyed a wide distribution at that time. Unsuccessful attempts to use extrusion for agglomeration of raw materials in the steel industry were made in the USSR in the 60s of the past century. The first successful experience of extrusion agglomeration of flue dust and sludge for a blast furnace took place at the Bethlehem Steel Company plant (in the United States) in 1993 [5]. A briquetting line with a capacity of 20 tons of raw briquettes per hour was commissioned. The equipment was supplied by J. C. Steele & Sons, Inc. (NC, USA). The line worked until the closure of the plant in 1996. In 1993, an extrusion agglomeration line was commissioned in Colombia at a ferronickel smelter. Gas cleaning and aspiration dusts from ferronickel production and laterite nickel ore fines were agglomerated without a binder. The capacity of the extrusion line is 700 thousand tons of briquettes annually. Currently, the plant is owned by BHP Billiton. Cost of production of one ton of briquettes does not exceed $2.1.

The lack of published results of a study of the metallurgical properties of these first extruded briquettes explains that the success of the first industrial projects on extrusion agglomeration did not attract wide attention. In the literature on briquetting in metallurgy, the technology of SVE for agglomerating metallurgical raw materials is either not mentioned [1] or the sphere of its application in metallurgy was limited to the manufacture of granules for sintering and its use for briquetting of metallurgical raw materials for blast furnaces was considered to be ineffective [36].

In 2010, the authors attempted to study the experience of the first industrial SVE lines. Based on the results of their operation and data posted on the official Internet sites of briquetting equipment manufacturers, including roller and vibropressing equipment, the process characteristics together with the properties of the briquettes were summarized in a comparative table (Table 1.7).

It is evident that SVE agglomeration technology is superior to existing industrial briquetting technologies in terms of its maximum capacity. None of the known roller and vibropressing briquetting factories have reached the scale of production attained by the aforementioned factory, owned by BHP Billiton in Colombia [37]. SVE requires a lower consumption of the cement Binder and is less demanding on the moisture content of the charge (drying is not required), in addition, it enables production of briquettes of a minimum size (diameter) compared to the size of the sinter and pellets.

Table 1.7 A comparison of the main parameters of briquetting technologies

The characteristics of the process and the properties of a briquette	Machines for briquetting and their characteristics		
	Vibropress	Roller-press	Extruder
Maximum capacity, ton/h	30	50	100
Maximum capacity of the industrial briquetting lines, tons per year	300	300	700
Cement binder content, %	8–10	15–16	3–9
Thermal processing of raw briquettes	80 °C (16–20 h)	–	–
Wastes generation	–	30% of production	–
Shape of briquette	Cylinder, prism	Pillow	Any
Dimensions, mm	+80 × 80	30 × 40 × 50	5–35
Moisture content of charge, %	<5%	<10%	8–15%
Possibility of immediate stacking of raw briquettes	–	Possible	Possible

1.6 Conclusions

In the nineteenth century, briquetting was the first and only industrial technology for agglomeration of fine anthropogenic and natural raw materials for the steel industry. Since the twentieth century, sintering and pelletizing have become the main agglomeration technologies. Manufacture of briquettes for use in iron, steel, and ferroalloys production processes using roller-pressing and vibropressing technologies currently enjoy limited application.

The metallurgical properties of vibropressed briquettes meet the requirements of the blast furnace process (except for their size) and they can be considered as components of a blast furnaces charge, however, a high cement binder content increases their costs and adversely affects the viscosity of the primary slag produced from these briquettes. The cost of such briquettes increases also as a result of the need to apply heat-moisture treatment since the mechanical strength of green briquettes is negligible.

Roller-press briquettes on organic binders are suitable for use in steelmaking units. The big disadvantage of roll-pressed briquetting is the high rate of returns (up to 30%) and the high mineral binder content.

At the end of the twentieth century, SVE agglomeration technology began to be used. The main advantages of the new technology compared with roller-pressing and vibropressing technology lie in their higher performance, reduced consumption of the binder, and the possibility of agglomerating wet materials.

SVE technology has a number of features that led to the emergence of a specific name for extruded briquettes—BREX.

The results of the operation of large-scale briquetting factories using SVE for agglomeration of laterite nickel ore fines and gas cleaning, and aspiration dusts from ferronickel production demonstrated the ability to achieve high levels of productivity from briquetting lines with low capital costs.

The results of the comparative analysis of technical and economic indexes of the three main industrial briquetting technologies enables SVE technology to be chosen rightly as the prospective technology for agglomeration of anthropogenic and natural raw materials in ferrous metallurgy.

References

1. Pietsch, W.: Agglomeration in Industry Occurrence and Applications, 375p. Wiley, New York (2005)
2. Yuzbashev, L.: Method for the preparation of artificial lumpy ore and artificial lumpy fuel from fines of ores and fossil fuels. Min. J. **2**(6), 257–288 (1901)
3. Blast Furnace Process.: Reference Guide. V.1. Moscow. State Scientific and Technical Publishing House of Literature on Ferrous and Non-Ferrous Metallurgy, 647p. (1963)
4. http://www.kuettner.com
5. Steele, R.B.: Agglomeration of steel mill by-products via auger extrusion. In: Proceedings of 23rd Biennial Conference, pp. 205–217. IBA, Seattle, WA, USA (1993)
6. BREX.: Certificate of trademark (service mark) No. 498006, application No. 2012706053 of 02.03.2012. Right holder A.M. Bizhanov
7. Koizumi, H., Yamaguchi, A., Doi, T., Noma, F.: Fundamental development of iron ore briquetting technology. Tetsu to Hagane **74**(6), 962 (1988)
8. Matsui, Y., Sawayama, M., Kasai, A., Yamagata, Y., Noma, F.: Reduction behavior of carbon composite iron ore hot briquette in shaft furnace and scope on blast furnace performance reinforcement. ISIJ Int. **43**, 1904 (2003)
9. Ravich, B.M.: Briquetting in Non-Ferrous and Ferrous Metallurgy, p. 232. Metallurgy, Moscow (1975)
10. Astakhov, A.G., Machkovskiy, A.I., Nikitin, A.I., et al.: Agglomerant's Reference Guide, p. 448. Technics, Kiev (1964)
11. Torok, J.: Briquetting for the steel industry: then and now. In: 32nd Biennale Conference of the Institute for Briquetting and Agglomeration, p. 20. Curran Associates (2013), 25–28 Sept 2011
12. Schwelberger, J., Brunner, C.H., Fleischanderl, A.: Recycling of ferrous by-products in the iron and steel. In: 34th Biennale Conference of the Institute for Briquetting and Agglomeration. Curran Associates (2013), 8–11 Nov 2015
13. Khazanova, T.P.: Production of manganese alloys from poor oxide and carbonate ores. In: Khazanova, T.P., Shirer, G.B., Lyakishev, N.P. (eds.) Development of Ferroalloy Industry in USSR, p. 122. Kiev (1961)
14. Khvichiya, A.P., Mazmishvili, S.M.: Silicomanganese smelting from ore briquettes in furnace with capacity of 16.5 MVA. Steel **2**, 138 (1970)
15. Sukhorukov, A.I., Sosedko, P.M., Khitrik, S.I.: Steel **2**, 135 (1970)
16. Mazmishvili, S.M.: Development and commercial exploitation of technologies for production of dust ore briquettes and smelting manganese ferroalloys from them. In: Mazmishvili, S.M., Simongulashvili, Z.A. (eds.) News of Higher Institutions, vol. 12, p. 43 (1992)
17. Guzman, I.Y. (ed.): Chemical Technology of Ceramics. Manual for Higher Institutions—Moscow: OOO RIF Stroimaterialy, 496p. Advertising and Publishing Firm Construction Materials, LLC (2003)
18. Kurunov, I.F., Bolshakova, O.G., Shcheglov, E.M., et al.: Metallurgist **6**, 36–39 (2007)

19. De Bruin, T., Sundqvist, L.: ICSTI/Ironmaking Conference Proceedings, pp. 1263–1273 (1998)
20. Paananen, T., Pisilä, E.: Improved raw material efficiency in hot metal production, pp. 1–8. METEC-ESTAD, Düsseldorf (2015)
21. Sharma, T., et al.: Effect of porosity on the swelling behaviour of iron ore pellets and briquettes. ISIJ Int. **31**, 312 (1991)
22. Sharma, T., et al.: Effect of reduction rate on the swelling behaviour of iron ore pellets. ISIJ Int. **32**, 812 (1992)
23. Sharma, T.: Swelling of iron ore pellets under non-isothermal condition. ISIJ Int. **32**, 960 (1994)
24. Singh, M., Bjorkman, B.: Cold bond agglomerates of iron and steel plant by-products as burden material for blast furnaces. In: Proceedings of REWAS 99: Global Symposium on Recycling, Waste Treatment and Clean Technology, San Sebastian, Spain, vol. 2, pp. 1539–1548 (1999)
25. Nicolle, R., Rist, A.: The mechanism of whisker growth in the reduction of wüstite. Metall. Trans. B **10**, 429 (1979)
26. Kempainen, A., Iljana, M., Heikkinen, E.-P., Paananen, T., Mattila, O., Fabritius, T.: ISIJ Int. **54**, 1539 (2014)
27. Titov, V.V., Murat, S.G., Kiselev, N.I.: The use of recycled materials in the charge of blast furnaces. Ecol. Ind. **1**, 16–21 (2007)
28. http://briket.ru/metallurg6.shtml
29. Kurunov, I.F., Kanaeva, O.G.: Briquetting is New Stage in Development of Technology for Sintering Raw Materials for Blast Furnaces, vol. 5, pp. 27–32. Bulletin of Scientific and Technical and Economic Information "Ferrous Metallurgy" (2005)
30. Kurunov, I.F., Shcheglov, E.M., Kononov, A.I., Bolshakova, O.G. et al.: Investigation of Metallurgical Properties of Briquettes from Technogenic and Natural Raw Materials and Assessment of Effectiveness of Their Application in Blast Furnace Smelting. Part 1, vol. 12, pp. 39–48. Bulletin of Scientific and Technical and Economic Information "Ferrous Metallurgy" (2007)
31. Kurunov, I.F., Shcheglov, E.M., Kononov, A.I., Bolshakova, O.G., et al.: Investigation of Metallurgical Properties of Briquettes from Technogenic and Natural Raw Materials and Assessment of Effectiveness of Their Application in Blast Furnace Smelting. Part 1, vol. 1, pp. 8–16. Bulletin of Scientific and Technical and Economic Information "Ferrous Metallurgy" (2008)
32. Kurunov, I.F., Malysheva, T.Y., Bolshakova, O.G.: Investigation of Phase Composition of Iron-Ore Briquettes in Order to Assess Their Behavior in Blast Furnace, vol. 10, pp. 41–46. Metallurgist (2007)
33. Sergeant, R., Onte, L., Huysse, K., et al.: Sergeant R. Heart management at Sidmar for an Optimal Hot Metal and Slag Evacuation, vol. 2. The 5th European Coke and Ironmaking Congress, Stockholm (2005)
34. http://www.kmz-tula.ru/articles-20100728.html
35. Tupkary, R.H., Tupkary, V.R.: An Introduction to Modern Iron Making. Khanna Publishers, Delhi (2013)
36. Ozhogin, V.V.: Foundations of Theory and Technology of Briquetting of Pulverized Metallurgical Raw Materials: Monograph, 442 p. PGTU, Mariupol (2010)
37. Bizhanov, A.M., Kurunov, I.F., Podgorodetskiy, G.S., et al.: Extrusion Briquettes (Brex) for Ferroalloys Production, vol. 12, p. 52. Metallurgist (2012)

Chapter 2
Features of Stiff Vacuum Extrusion as a Method of Briquetting Natural and Anthropogenic Raw Materials

Stiff vacuum extrusion (SVE) technology is applied in the production of ceramic bricks in 64 countries around the world, including the United States, Britain, Germany, South Korea, and South Africa. The world's largest brick factory in Saudi Arabia produces a million bricks per day using SVE technology [1].

In accordance with brick industry terminology, the word "stiff" is used to describe the process of extrusion, which is carried out at pressures ranging from 2.5 to 4.5 MPa and moisture contents ranging from 12 to 18% (Table 2.1 [2]).

2.1 Preparation of Burden Materials for SVE Briquetting

One of the basic criteria determining the suitability of extrusion material is its plasticity—a trait that assures that it can be effectively pushed through the holes in die. A necessary condition for the status of plasticity is complying with the granulometric composition and moisture content requirements for SVE. In some cases, additional pulverization may be required in order to achieve the desired plasticity of material.

Unlike a roller press and vibropress briquetting, shear stress plays an important role in SVE agglomeration. Shear stress occurs when the mixture is processed in the screw feeder, in pug mills, and then in the extruder. In [3], based on a comparison of coal briquette porosity values in various pressing options (compression and its combination with torsion), it was found that more dense briquettes (less porous) are formed in the combined pressing option (under identical values of applied pressure). With full compression, a significant proportion of energy is expended on the elastic deformation of the particles themselves, while, in the presence of shear stress, the convergence of particles on the surface forces activation distance is more effective. In full compression of close-packed particles, each particle only comes into contact with its immediate neighbors and is subjected to a compression load. Under shear stress, the particles of the adjacent layers are subjected to abrasion due

© Springer International Publishing AG 2018

I. Kurunov and A. Bizhanov, *Stiff Extrusion Briquetting in Metallurgy*,
Topics in Mining, Metallurgy and Materials Engineering,
https://doi.org/10.1007/978-3-319-72712-7_2

Table 2.1 Types of extrusion

Type of extrusion	Low-pressure extrusion	Medium-pressure extrusion	High-pressure extrusion	
Designation used in structural ceramic industry	Soft extrusion	Semi-stiff extrusion	Stiff extrusion	
Extrusion moisture, % on dry	10–27	15–22	12–18	10–15
Extrusion pressure, MPa	0.4–1.2	1.5–2.2	2.5–4.5	Up to 30

to contact with irregular surfaces, which can lead to crushing, the opening up of new surfaces, and, hence, to an increase in the number of contacts between particles of the mixture.

In order to identify the possible impact of shear stress on the change of particle size distribution of briquetted material in conditions of SVE, we have compared the results of pulverization of coke fines in three different ways: in a hammer mill, in a roller crusher, and by double shearing through shearing plate in the extruder (Fig. 2.1). Samples of original coke breeze had the following properties: 8.97% moisture content; 13.28% ash content; volatile content 3.45%; 0.61% sulfur content. Since coke breeze particle was larger in size than what is usually required for extrusion, they were pulverized.

Fig. 2.1 Equipment and tools for materials pulverization (top left—hammer mill; top right-roller crusher; bottom left—shearing through the shearing plate of extruder; bottom right—shearing plate)

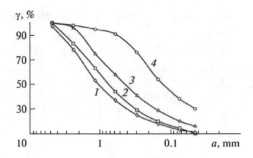

Fig. 2.2 Granulometric composition of coke breeze in the following states: (1) initial and (2–4) after additional grinding in a hammer mill, in a roll crusher, and double extrusion in an extruder, respectively. γ is the yield of the oversize mass, and a is the mesh size

Wet screens were used to determine particle size distributions of raw materials. Moisture content was measured using a moisture balance. A calibrated electronic scale with a density measuring attachment was used to determine pellet density. The results of the granulometric analysis are presented in Fig. 2.2.

As evidenced above, the grinding of coke breeze is maximized after double extrusion through a shearing plate in an extruder. In this case, the effect of deep grinding is achieved through the application of high shear stresses. The use of a hammer mill for such a material was found to be ineffective, and the granulometric composition of the ground material differed weakly from that of the initial coke breeze.

Based on these results, we decided to study the effects of the grinding method on the strength of the brex produced from coke breeze pulverized in different manners. Three portions of brex of the same composition—94% of coke breeze, 5% of Portland cement, and 1% bentonite—were manufactured. The only difference between the brex was the manner in which the coke breeze was ground. The brex were classified assigned the following numbers in accordance with the method of grinding: No. 1 was coke breeze ground by roller crusher; No. 2 was coke breeze sheared twice through the shearing plate with a plurality of holes; No. 3 was coke breeze ground by hammer mill. Extrusion parameters and physical properties of brex are shown in Table 2.2.

It follows from these data that the extrusion of the first two mixtures was carried out in accordance with similar processing parameters and that the extrusion of the coarser particles in mixture No. 3 (hammer mill) is accompanied by an increase in the temperature of the material. As a result, the brex made of the largest particles turned out to have the minimum strength in axial compression tests. As for energy consumption, the extrusion of mixture No. 1 was most efficient. Its excellent extrusion ability can be attributed to the material particle shape after grinding in a roll crusher, which favors the plane-parallel orientation of particles.

The difference between the tensile strengths of samples No. 1 and No. 2 is insignificant and only indicates an earlier beginning of cracking in brex No. 2. It also follows from this data that the density of brex after double extrusion exceeds

Table 2.2 Extrusion parameters and physical properties of coke breeze brex

Sample of brex (grinding method)	Moisture content, %	Temperature °C	Vacuum, mm Hg.	Density, g/cm³	Compressive strength, kgf/cm²
No. 1 (roller crusher)	16.5	30.56	15.24	1.63	37.76
No. 2 (double shearing)	16.7	33.33	17.78	1.67	34.32
No. 3 (hammer mill)	16.6	55.56	81.28	1.63	20.25

that of brex samples made of ground and milled coke breeze by 2.5%. Obviously, the dense packing of brex No. 2 particles is the result of a high degree of material grinding. In contrast to brex samples No. 1 and No. 3, no mixture dewatering was detected during extrusion in this case. Differences in compressive strengths of brex samples prepared from differently treated coke breeze of the same batch—can result from a number of factors related to different particle sizes, shapes, and surface relief. The shape of a particle after grinding depends on material characteristics and the grinding method, including time [4–6]. For example, it is generally accepted that, after grinding in roll crushers, the material mainly consists of angular particles, whereas the material particles usually maintain the same size and a rounded shape after ball and hammer milling. In roll crushers, grinding occurs under the compressive, shear, and rubbing forces. As a result, you have rough particles with sharp projections, many edges and corners, and (correspondingly) a large contact surface form. When the material is ground by hammer mill, the particle surfaces are polished by way of impact and the particles acquire a rounded shape.

For example, the researchers [7] studied the effect of grinding device on oil coke particle shape of (1) low-porosity coke with thick cell walls without visible cracks and (2) fissured coke whose porosity and cell wall thickness of which were distributed over a wide range. In the first instance, grinding method did not affect the shape of particles 200–600 μm in size. In the second instance, it was revealed that the particle shape is dependent on particle size in various ways when ground in a hammer mill and in a roll crusher (Fig. 2.3) [7].

These curves reflect the following dependence of the shape factor: $\phi = 1.1 V1/3N1/6A-1/2$, where V is the specific particle volume (cm³/g), N is the number of particles in 1 g substance, and A is the specific surface area (cm²/g). The material particles ground in a hammer mill are characterized by stable high shape factors, which indicate the closeness of most of the particles' forms to a rounded shape in

Fig. 2.3 Particle shape factor ϕ versus the average particle size upon grinding in (1) hammer mill and (2) roll crusher [7]

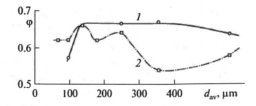

Fig. 2.4 Structure of coke breeze particle (scanning electron microscopy)

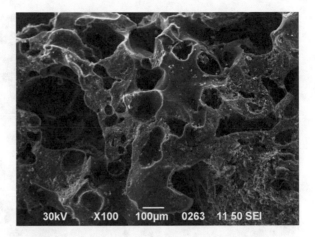

the indicated size range. For the particles of the material ground in a roll crusher, the shape factors in the 140–600 μm particle size range are lower and minimal at a size of ∼350 μm, which supports the conclusion about a more nonuniform particle shape after grinding in a roll crusher.

As in [7], we used a porous and fissured material. Figure 2.4 shows an image of coke breeze particles taken with a JEOL JSM_6490 LV scanning electron microscope (SEM). In Fig. 2.5, we compare the particle shapes of the initial coke breeze and the coke breeze after additional grinding. The initial coke breeze particles are characterized by hillocks (rounded projections), which are typical of coke; they belong to internal (not opened) pores located under the upper layers of the carbon material. It is seen that the particles after a roll crusher and double extrusion through an extruder have a nonuniform angular shape, whereas the coke breeze particles ground in a hammer mill have a rounded shape.

The brex samples were subjected to tensile splitting tests on a bench-type one-column electromechanical Instron 3345 tensile testing machine at a load of 5 kN. When studying the statistics of brex orientation distribution in a charge, we [8, 9] found that this type of external load is most probable for a cylindrical brex. Figure 2.6 shows the results of testing specially prepared cylindrical specimens of brex 1–3 25 mm in diameter and 20 mm in height. It is seen that, at approximately the same carrying ability, the brex specimens' reactions to an applied load are different.

The difference in the maximum loads can be related to defects in the specimens. However, the difference in the characters of behavior can have radically different causes. Brex No. 2 demonstrates ductile fracture, which is indicated by the existence of a yield plateau, i.e., the horizontal component of the brex No. 2 curve. This phenomenon is thought to be explained by a "relay-race" grain to grain gliding transfer in accordance with the Hall–Petch equation stipulating an inverse relationship between grain size and yield strength [10]. In this case, a grain boundary is a barrier to dislocation motion, which causes dislocation nucleation and development

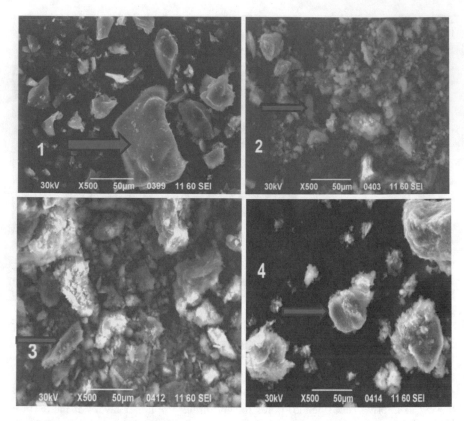

Fig. 2.5 Micrographs of (1) initial coke breeze particles and after additional grinding, (2) in roll crusher, (3) by double extrusion through an extruder, and (4) in a hammer mill

Fig. 2.6 Load P-displacement Δl curves for cleavage tensile tests of brex

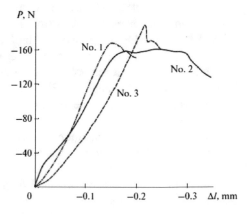

in a neighboring grain. In other words, the larger the number of barriers to be overcome, the lower the dislocation motions dynamics and the higher the crack development resistance.

Apparently, there exists a threshold particle size, below which cracks cannot propagate in brex. Obviously, the granulometric composition of the brex No. 2 mixture favors this scenario due to the highest content of thin particles among the other mixtures (see Table 2.2). The integrity of brex No. 2 is retained even after tests; therefore, high impact strength can be expected. At this type of loading, local fracture zones will not cause full fracture and debris formation in a specimen. The load–displacement curves of brex No. 1 and No. 3 decrease significantly after crack initiation, which points to more brittle fracture of these specimens. The brittle fracture in brex No. 1 develops more slowly than it does in brex No. 3 because of a smaller average particle size. Note that these circumstances are very important for briquetting. Several stages of charging–discharging related to throwing down briquettes can be required to supply the briquettes to a furnace even in one enterprise. When the impact toughness of an agglomerated product is increased, the logistics of supplying it to the site, including those that are far away, can be significantly simplified and, hence, be made cheaper. It is important for stiff extrusion that a change in the behavior of brex under external mechanical action was achieved using same agglomeration equipment.

In addition to the physical mechanisms that weaken dislocation propagation, leading to a high strength of brex made of small particles, the distribution of a binder in the brex body also substantially contributes to its strength. It is clear that the low strength of brex No. 3 during compression can be explained by a small amount of contacts between particles. Figure 2.7 shows micrographs of the particle surfaces in brex No. 3 (grinding in a hammer mill) and brex No. 2 (double extrusion through a shearing plate). It is clearly visible that, because of the lower particle surface roughness of brex No. 3 as compared to brex No. 2, the number of hillocks with a binder (cement and bentonite) on their surface (bright aggregates) is significantly lower. (Since minerals contain heavy chemical elements, they manifest themselves as bright precipitates in micrographs.) Therefore, the surface relief of the brex No. 2 particles also favors good adhesion of particles due to the large binder volume that covers the sites of particle–particle contacts.

When studying the particle surfaces of brex No. 2, we detected bentonite fibers, which were described for the first time in [11]. Figure 2.8 shows the surface of brex No. 2 particle and the surfaces of glass microspheres subjected to soft rolling (without crashing) in a roll crusher which imitated shear stresses. This treatment promoted the development of bentonite fibers covering the particle surface. After this preliminary treatment of a magnetite concentrate and bentonite mixture, the fraction of bentonite required for the given strength decreased twofold (from 0.66 to 0.33% concentrate mass). The appearance of this structure in brex No. 2 is likely to result from the shear stress applied to the material during auger extrusion. The effect of the appearance of bentonite fibers is less pronounced in brex No. 1 and No. 3, which is likely to be associated with a lower fraction of thin particles. Brex No. 1 and No. 3 have predominantly lamellar and flake-like bentonite particles (Fig. 2.9).

Fig. 2.7 SEM micrographs of the particle surfaces in brex No. 3 (**a**) and No. 2 (**b**)

A mixture becomes suitable for extrusions through the addition of a plasticizing agent, mainly bentonite, to a charge. The fraction of bentonite required to ensure the extrusion of a mixture oscillates in the 0.25–1.0% range of the briquetted material density. An important consequence of the plasticized mixture is an increase in the mechanical strength of brex and a change in the character of its fracture from brittle to viscoplastic fracture.

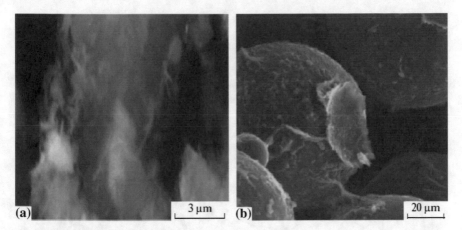

Fig. 2.8 Bentonite fibers on **a** the particle surface in brex No. 2 and **b** on the surface of glass microspheres [11]

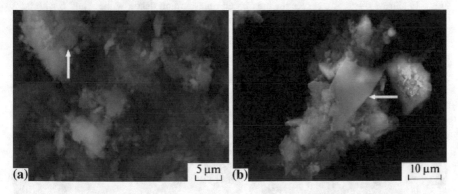

Fig. 2.9 Bentonite particles in brex No. 1 (**a**) and No. 3 (**b**)

Below, we present axial compression test results of first series brex made of a charge of a base composition of 67% dust of the aspiration of ferrochrome production, 15% chromium ore concentrate, 15% coal, 3% Portland cement, and second series brex made of the same charge with an addition of 0.5% bentonite that were performed in a laboratory extruder. The axial compression tests of the brex of both series were carried out on a bench-type one-column electromechanical Instron 3345 tensile testing machine at a load of up to 5 kN. Figure 2.10 shows the stress–strain curves that demonstrate the change in the character of fracture of bentonite-containing brex from pronounced brittle to viscoplastic fracture. This change is also illustrated in Fig. 2.11, which shows that in the case of fracture of bentonite-containing brex, the compression surfaces of the tensile testing machine should be much closer than they are in the case of bentonite-free brex. A comparison of the strengths of the brex of two series showed that the average

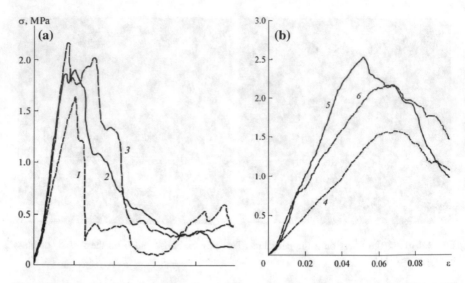

Fig. 2.10 Stress σ–strain ε curves for brex based on aspiration dust **a** without and **b** with 0.5% bentonite

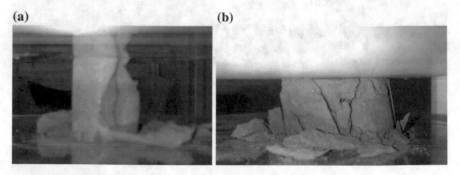

Fig. 2.11 Fracture of brex based on aspiration dust: compositions **a** without and **b** with 0.5% bentonite. The initial brex sizes are 50 mm in height and 25 mm in diameter

ultimate compressive strength of the brex made of a bentonite-free mixture is 1.88 MPa and that of bentonite-containing brex is 2.08 MPa (higher by 10%).

As for traditional methods of briquetting, the favorable effects of bentonite or bentonite in combination with Portland cement on the strength of briquettes is well known [12, 13]. An attempt to determine the optimum ratio of the fractions of Portland cement and bentonite in a charge to be briquetted was described in patent [14]. It was stated that the addition of a bentonite (15–25%) and Portland cement (85–75%) binder enhances the mechanical strength of the corresponding briquette. For briquettes consisting of converter sludge (24.3%), scale (23.4%), blast furnace dust (23.4%), iron ore fines (18.9%), and binder (10%) consisting of 75% Portland cement and 25% bentonite, the result of a drum test (fraction of a size of >6.3 mm) increased to 76% (it was 67% for briquettes of the same composition but without bentonite).

We studied the influence of this combined binder on both the compressive strength of brex and the rate of strength augmentation during strengthening storage under natural conditions. For tests, we chose the brex that were produced in the industrial line of stiff vacuum extrusion located at Suraj Products Ltd. (India) and contained (%) 47.2 converter sludge, 28.3 blast furnace dust, 18.9 iron ore fines, 4.7 Portland cement, and 0.9 bentonite. Portland cement and bentonite were manually mixed in a dried state and added to a charge mixture before a pug sealer. For brex samples, we measured compressive strength σ (on Tonipact 3000 (Germany) according to standard DIN 51067), open (apparent) porosity η (vacuum method of liquid saturation according to standard DIN 51056), and density ρ (on Metler (United States) balance) daily. Table 2.3 shows the results of our daily measurements over the course of 9 days.

Figure 2.12 shows the brex porosity and strength curves measured during structuring storage. A pronounced local maximum is clearly visible in the brex strength curve on the third day. On the next day, this changes into softening. The strength increases over the course of further storage. Note that, before softening, the brex strength accounts for $\sim 84\%$ of the brex strength after strengthening storage for 1 week. The change of open porosity almost repeats the change of strength, except for the first day of strengthening. The decrease in the porosity at that time is obviously related to the swelling of bentonite, which fills the pore space [15].

We obtained similar results for brex strength during compressive strength and splitting strength tests performed on a strength-testing machine consisting of a hand-power press, a strain gage, and a recording device (Fig. 2.13).

On the third day of strengthening, the brex samples retained their integrity through the end of testing and demonstrated a viscoplastic character of fracture during compressive and tensile splitting tests. On the third day of strengthening, both halves of the brex were connected despite the development of almost a

Table 2.3 Results of the daily measurements of the brex apparent porosity, density, and compressive strength

Maturing time, days	Apparent porosity %	Density (g/cm^3)	Compressive strength (kg/cm^2)
1	31.5	2.42	24
2	25.4	2.66	45
3	32	2.43	63
4	27	2.44	52
5	27.2	2.45	56
6	26.2	2.45	57
7	26.8	2.46	59
8			75
9			80

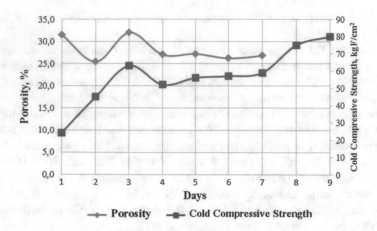

Fig. 2.12 Changes in compressive strength and porosity of brex during strengthening 9-day storage

Fig. 2.13 a Axial compression and **b** tensile splitting tests of the brex produced at Suraj Products Ltd. at the third (**a, b** on the left) and seventh (on the right) day of strengthening storage. The initial brex sizes are 25 mm in height and 25 mm in diameter

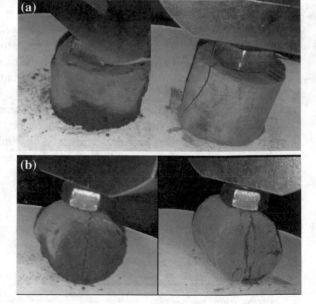

complete crack in the tensile splitting test. The compression and splitting of brex led to the release of moisture on the bottom. Moisture release was absent during compression on the seventh day of strengthening. When samples were subjected to compressive and tensile splitting tests on the seventh day of strengthening, the viscoplastic character of fracture was substantially lost.

This effect of nonmonotonic strengthening of briquettes with a cement binder, characterized by a local strength maximum and accompanied by viscoplastic behavior at a breaking load, had been neither detected nor described previously. The strengthening of monomineral cement stone and concrete is known to be of a monotonic character [16]. This nonmonotonic character is well noted in the process of increasing the strength of binderless brex or of brex with an alternative (no cement) binder. Ozhogin [17] presents data on the impact fracture (analog of dropping strength) oscillations of briquettes having no binder and having different compositions upon drying under natural conditions (at a temperature of 20 °C). The detected softening was thought to be caused by recrystallization, and it was recommended to limit the transportation and transfer of briquettes at this time and to store briquettes in closed vessels.

The application of a cement–bentonite binder changes the behavior of brex substantially during a braking action, increases its impact strength, and decreases the probability of brittle fracture.

This behavior of brex can be explained by cement–bentonite binder properties related to the formation of coagulation structures in the cement–bentonite–water system, which leads to the modification of the properties of the binder. Such structures are known to form in the gel–cement solutions used for the cementation of boreholes. The properties of the gel–cement solutions and the theoretical and practical aspects of the formation and fracture of cement–bentonite systems are well known and systematized in, e.g., [18]. The driving force of the formation of such structures is the attraction of negatively charged bentonite particles to positively charged Portland cement particles, which results in their rapid coagulation and the formation of a suspended cement particle structure. Hydrated cement particles are gradually coated with an impermeable shell of flaky bentonite particles. The number of adsorbed bentonite particles is proportional to the activity of cement. During hydration, Portland cement particles grow in size, which leads to tension, a break in the integrity of bentonite shells, and the penetration of water to cement particles (i.e., to further hydration of cement and, apparently, the adsorption of a larger amount of bentonite).

The fracture intensity of such a coagulation structure depends on the cement activity. The application of slag Portland cement is likely to retard this process. The coagulation structure begins to fail gradually as a result of the coagulation effect of calcium ions and changes into the structure of a hardening cement stone. The decrease in the strength after the peak at the third day is explained by the decomposition of the coagulation structure (Fig. 2.12). The further increase in strength is completely determined by the hydration of cement stone.

The possibility of such a stable structure forming is supported by the conclusions of work [19]. The range of the relative content of bentonite and Portland cement in the binder used for the production of brex (80% Portland cement, 20% bentonite) corresponds to the dark gray region in the water–cement–bentonite phase diagram (Fig. 2.14 [19]).

Note that the local moisture content near coagulated particles exceeds the moisture content of the as-prepared brex substantially (by 11–12%) due to differences in the permeability of minerals. An increase in the fraction of bentonite in the binder up to 30–40% favors the formation of a stable mastic suspension.

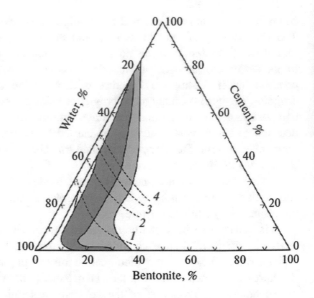

Fig. 2.14 Water–cement–bentonite phase diagram [19]. (Light gray) region of formation of a stable mastic suspension and (dark gray) region of a stable and mobile suspension. (1–4) compressive strength values, kPa: (1) 69, (2) 345, (3) 689, and (4) 1378

The effect revealed in this work can be important for practice: brex based on a cement–bentonite binder can be used as a charge component for a metallurgical furnace within 3 days of drying under natural conditions. As a result, the required sizes of brex storage can be decreased. The threshold level of compressive strength can be easily achieved by a simple increase in the fraction of this combined binder. Note that the total fraction of a binder in the brex produced at Suraj Products Ltd. accounts for at most 5.6% of the brex mass, which is significantly lower than the required content of Portland cement used for vibropressing briquetting (10–12%). The use of bentonite in the binder composition changes the character of fracture of briquettes and makes it viscoplastic, which simplifies the transportation of brex to the site of its application. A quantitative description of the detected effect is the subject of further investigation.

The moisture release during the briquette strength tests on the third day can be related to restructuring in the agglomerates formed in cement during its hydration, which is accompanied by the release of part of chemically fixed water. It is known that, at a sufficiently high content of volume capillary moisture in cement, intense nucleation and growth of acicular aggregates take place on the third day of hardening [20]. It is these aggregates that are responsible for the connection of cement grains and prevent crack development in primary aggregates, and these aggregates were experimentally detected in the cleavages of commercial brex. The experimental investigation was carried out using a high-resolution scanning electron microscopy (SEM) on an Auriga CrossBeam (Carl Zeiss) analytical working station equipped with an INCA X-Max energy dispersive spectrometer. The accelerating voltage was 20 kV. During the hydration of cement, ettringite forms as a result of the reaction between calcium aluminate and sulfate: $3CaO \cdot Al_2O_3 + 3CaSO_4 \rightarrow$ Ettringite.

Fig. 2.15 SEM micrograph of acicular aggregates (ettringite particles) surrounded by bentonite plates in cement in the structure of the brex produced at Suraj Products Ltd.

Using SEM, we were able to reveal the elements of the decomposed coagulation structure. Ettringite particles surrounded by bentonite plates are visible in Fig. 2.15.

The change in the properties of brex induced by the addition of bentonite to the composition of the mixture to be briquetted also manifests itself after the preliminary homogenization of the mixture to be briquetted during its "souring" (storage of the moisturized mixture with added bentonite for within a certain time). This method allows achieving a high degree of homogeneity of mixture properties for subsequent briquetting. In some cases, the mechanical strength of brex made of homogenized charge can significantly increase. Consumption of plasticizer for briquetting the charge without prior homogenization, in this case, can decrease significantly. We studied the effect of souring on the strength of brex of various compositions (Table 2.4). Table 2.5 gives the results of tensile splitting tests of brex 1 week after its manufacture. These results demonstrate that the use of bentonite in combination with Portland cement improves the strength properties of brex. The strength of brex increases substantially in some cases. Furthermore, the manner of their fracture also changes and signs of viscoplastic fracture appear. In this case, brex can better withstand the impact loads that appear during their transportation to the sites of their application.

Table 2.4 Brex compositions for testing of the souring efficiency

Brex composition/No.	1	2	3	4	5	6
BF sludge	42.8	41.8	41.8	41.2	47.8	28.3
BOF sludge	39.8	38.8	38.8	38.2	43.7	25.4
Iron ore concentrate						29.3
Mill scale	13.0	13.0	13.0	12.1		10.7
Portland cement	4.0	4.0	6.0	8.0	8.0	5.8
Bentonite	0.4	0.4	0.4	0.5	0.5	0.5
Microsilica		2.0				

Table 2.5 Values of tensile splitting strength of brex

Tensile splitting strength of brex, MPa	1	2	3	4	5	6
Without souring	0.86	1.93	2.08	1.00	1.01	0.77
After souring	2.45	3.83	5.76	1.88	1.29	1.26
The ratio of strength values	2.85	1.98	2.76	1.88	1.28	1.64

Table 2.6 Granulometric composition of brex components

Material	Fraction, mm/yield, mass%					
	$-20... + 10$	$-10... + 5$	$-5... + 2$	$-2... + 1$	$-1... + 0.5$	-0.5
Manganese ore concentrate, %	3.42	25.42	32.58	18.95	9.70	9.93
Baghouse dust of SiMn production	Fine dust, size less than 0.063 mm					

The results of the study on the effects of shearing stress on the properties of extruded mixture allow using shearing through the shearing plate for the homogenization of the mixture prior to its agglomeration with binder.

To demonstrate the effect of such operations on the homogenization of the mixture, we have compared values of compressive strength on brex composed of manganese ore concentrate with added baghouse dust resulting from silicomanganese production with and without homogenizing the earlier-sheared mixture.

The granulometric composition of brex components is listed in Table 2.6. The composition of brex is given in Table 2.7.

A laboratory extruder has been used for brex sample production. The sheared mix was subjected to the souring over the course of 4 h. Table 2.8 shows the results of the measurement of the compressive strengths of brex No. 1–3 produced from this mix with and without souring. The increase in compressive strength a week after manufacturing was 14.2% for brex No. 1; 7.62% for brex No. 2, and 54.5% for brex No. 3.

Study of the brex structure by means of scanning electronic microscopy revealed differences in the structure and distribution of pores. Figure 2.16 presents the structure of the brex No. 2.

One can see that pore size is significantly smaller in the brex made from a sheared and soured mix. Our results confirm the efficiency of using a shearing extruder in preparing the briquette charge for souring.

Table 2.7 Composition of brex

Composition of brex	No. 1	No. 2	No. 3
Manganese ore concentrate	80	66	56
Baghouse dust	14	28	38
Portland cement	5	5	5
Bentonite	1	1	1

Table 2.8 Compressive strength of brex No. 1–3

Strength/ brex No	No. 1	No. 1— soured	No. 2	No. 2— soured	No. 3	No. 3— soured
Day 1 (MPa)	2.9647	2.7579	2.6890	5.7571	4.0955	6.1708
Day 3 (MPa)	4.8608	4.8608	6.0674	7.8255	6.1019	10.2042
Day 7 (MPa)	6.6741	7.6187	13.1345	14.3411	10.1698	15.7200

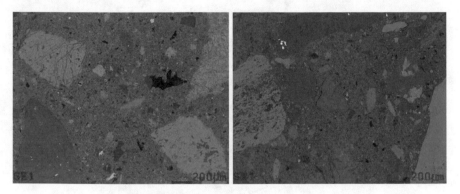

Fig. 2.16 Scanning electron microscopy of the brex No. 2 structure. Left—without souring; right—with souring during 4 h after shearing

2.2   Technological Process of Briquetting by Method of Stiff Vacuum Extrusion

Typical layouts of the SVE briquetting line is shown in Fig. 2.17.

Prepared in the warehouse of the raw material mixture of the main components of the brex mix is being fed by an even feeder (Fig. 2.18), equipped with spirals cast of chrome alloy.

Next, the prepared mixture with added binder and plasticizer is fed for mixing in the pug sealer. The line can also contain primary open pug-mill. The pug sealer consists of a large open part and the sealing node. The open part (Fig. 2.19) consists of a trough and system blades for mixing. The blades are fastened to the steel rod shaft by bolted clamps, making it possible to rotate the blades to adjust the angle at which the processing takes place and, thereby, change the machine's performance.

The pug sealer is combined in a single unit with the extruder and is positioned above it (Fig. 2.20).

The mixture enters the vacuum chamber partially agglomerated (Fig. 2.21) and due to the high vacuum inside the chamber and removal of air and moisture, the pieces of the mixture immediately crumble into isolated particles, which fall down on the blades of the auger. It is known [21] that the air adsorbed by the surface of the particles of plastic material in the form of multi-molecular layers held by van der Waals forces slow the rate at which they wetten, prevents the mass's uniform

BATCH MIXER

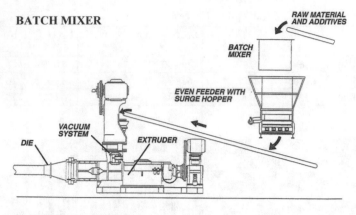

CONTINUOUS MIXING

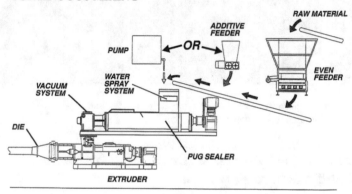

Fig. 2.17 Typical layouts of SVE briquetting line

Fig. 2.18 J.C. Steele E
Series Even Feeder

Fig. 2.19 Mixing in the open part of the pug sealer

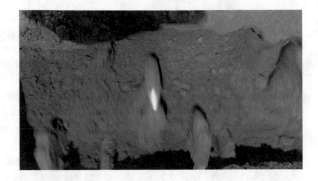

Fig. 2.20 The appearance of extruder (bottom) and pug sealer

Fig. 2.21 Partial agglomeration of the mix in pug sealer

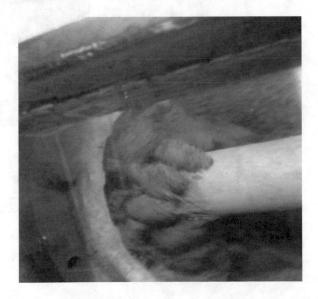

compaction, and promotes elastic deformation, which results in lamination and micro-cracks detected during the drying and firing of products. In filling the pores, the air also prevents moisture from penetrating, separates the particles of the mass, and acts as a leaner. The vacuum removes air from the pores and promotes the closer contact of particles.

The vacuum is maintained throughout the working volume of the extruder up to the die. The pressure of the vacuum is at least 100 mm Hg (in absolute value). The area of the vacuum in the working chamber of the extruder and pug sealer is shown in Fig. 2.22. The combination of mechanical pressure and vacuum in the working extruder chamber helps to remove almost all compressible air from the material before densifying, which leads the green brex to have high values of compressive strength of and to immediately be transported by conveyors and stockpiled, practically without fines. In addition, as is well known, the vacuum slightly decreases the viscosity of the cement paste, which facilitates its uniform distribution in the briquetting mass and improves its interaction with water [22]. This, combined with a higher density of mass resulting from the removal of air, leads to a decrease in cement binder consumption.

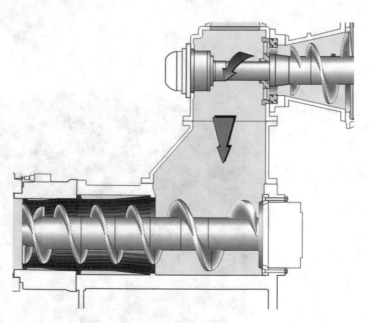

Fig. 2.22 Vacuum area in extruder and pug sealer

2.3   The Movement of the Briquetted Mass in the Extruder

Due to rotation of the auger blades in the working chamber of the extruder, formable mass performs translational and rotational motion, which is slowed by the walls (Fig. 2.23).

In the conveying zone, the material is loose and moves along the barrel without densification. Bulk density remains unchanged. Zone 2 is the densifying region where the loose material is compacted. In zone 3, metering is achieved by way of the special geometry of the wings of the point auger. Zone 4 serves the purpose of distributing the pressure generated by the metering zone more evenly over the die, thus tending to yield a more even flow through it.

Bizhanov et al. made a preliminary assessment of the briquetted mass movement in the working chamber of the extruder from the point of view of basic continuum mechanics [23]. The rotating mass of a wet, continuous medium has an isotropic molecular pressure field [2, 24] and is subject to the fundamental laws of conservation of mass, momentum, and energy [25–27] with an appropriate rheology for the coefficient of dynamic viscosity μ [28–30].

First of all, as in [31], the motion of a particle, or point $\mathbf{r} = x\mathbf{i} + y\mathbf{j} + z\mathbf{k}$, will be measured in cylindrical coordinates r, θ, z,

$$r = \left(x^2 + y^2\right)^{1/2}], \text{ and } \theta = \text{acrtg}\frac{y}{x}, \text{ or } x = r\cos\theta \text{ and } y = r\sin\theta,$$

or in polar orts

$$\mathbf{I} = \cos\theta = \mathbf{i}\cos\theta + \mathbf{j}\sin\theta \text{ and } \mathbf{J} = -\sin\theta = \mathbf{j}\cos\theta - \mathbf{i}\sin\theta \text{ and } \mathbf{k}$$

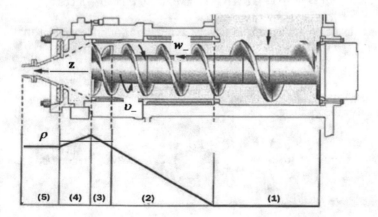

Fig. 2.23 Stages of densifying in the working zone of the extruder. 1—conveying, 2—densifying, 3—metering, 4—pressure distributing, 5—die

preserving the sum of the double orts of the direct product and

$$\mathbf{ii} + \mathbf{jj} = \mathbf{II} = \mathbf{JJ},$$

"inheriting" the properties of the sine and cosine functions,

$$\mathbf{I}_\theta = \partial_\theta \mathbf{I} = \frac{\partial \mathbf{cos}\,\theta}{\partial \theta} = -\mathbf{sin}\,\theta = \mathbf{J} \text{ and } \mathbf{J}_\theta = -\partial_\theta \mathbf{sin}\,\theta = -\mathbf{cos}\,\theta = -\mathbf{I},$$

but otherwise remaining the same as the Cartesian orts $\mathbf{i}, \mathbf{j}, \mathbf{k}$:

$$\mathbf{I} \times \mathbf{J} = \mathbf{i} \times \mathbf{j} = \mathbf{k}, \ \mathbf{J} \times \mathbf{k} = \mathbf{I}, \ \mathbf{k} \times \mathbf{I} = \mathbf{J},$$
$$\mathbf{I} \cdot \mathbf{J} = \mathbf{J} \cdot \mathbf{k} = \mathbf{k} \cdot \mathbf{I} = 0 \text{ and } \mathbf{I} \cdot \mathbf{I} = \mathbf{J} \cdot \mathbf{J} = \mathbf{k} \cdot \mathbf{I} = 1,$$

so:

$$\mathbf{r} = x\mathbf{i} + y\mathbf{j} + z\mathbf{k} = r\mathbf{I}(\theta) + z\mathbf{k}.$$

In addition the specified field mass accelerations,

$$\mathbf{g}(t, \mathbf{r}) = g^x\mathbf{i} + g^y\mathbf{j} + g^z\mathbf{k} = g^r\mathbf{I} + g^\theta\mathbf{J} + g^z\mathbf{k},$$

field velocities or flow

$$\mathbf{u}(t, \mathbf{r}) = u^x\mathbf{i} + u^y\mathbf{j} + u^z\mathbf{k} = u\mathbf{I} + v\mathbf{J} + w\mathbf{k} = \mathbf{r}_t,$$

receives three components of velocity, axial $w = z_t$, radial, $u = r_t$, and azimuthal, $v = r\theta_t$ from the polar orts

$$u^z = w, u^x = u \cos\theta - v \sin\theta, \ u^z = u \sin\theta + v \cos\theta,$$

will be considered further to be axially symmetric, or independent of the polar angle θ:

$$u_\theta = v_\theta = w_\theta = \rho_\theta = p_\theta = g^r_\theta = g^z_\theta = 0 \text{ for } g^\theta = 0.$$

In this case, the gradient

$$\nabla = \begin{pmatrix} \mathbf{i}\partial_x \\ +\mathbf{j}\partial_y \\ +\mathbf{k}\partial_z \end{pmatrix} = \begin{pmatrix} \mathbf{I}\partial_r + \mathbf{k}\partial_z \\ +\mathbf{J}\dfrac{\partial_\theta}{r} \end{pmatrix}, \partial_x = \cos\theta\partial_r - \frac{\sin\theta}{r}\partial_\theta, \partial_y$$
$$= \sin\theta\partial_r + \frac{\cos\theta}{r}\partial_\theta,$$

and divergences

$$\nabla \cdot \rho \mathbf{u} = \begin{pmatrix} \mathbf{I}\partial_r + \mathbf{k}\partial_z \\ +\mathbf{J}\dfrac{\partial_\theta}{r} \end{pmatrix} \cdot \begin{pmatrix} \mathbf{I}\rho v \\ \mathbf{J}\rho v \\ +\mathbf{k}\rho w \end{pmatrix} = \begin{pmatrix} (\rho u)_r \\ +(\rho w)_z \end{pmatrix} + \dfrac{1}{r}\mathbf{J} \cdot \begin{pmatrix} \rho u \mathbf{I}_\theta \\ +\rho v \mathbf{J}_\theta \end{pmatrix}, \mathbf{I}_\theta$$

$$= \mathbf{J}, \mathbf{J}_\theta = -\mathbf{I},$$

and

$$\nabla \cdot \mu \nabla u = \begin{pmatrix} \mathbf{I}\partial_r + \mathbf{k}\partial_z \\ +\mathbf{J}\dfrac{\partial_\theta}{r} \end{pmatrix} \cdot \begin{pmatrix} \mathbf{I}\mu u_r \\ +\mathbf{k}\mu u_r \end{pmatrix} = \begin{pmatrix} (\mu u_r)_r \\ +(\mu u_z)_z \end{pmatrix} + \dfrac{1}{r}\mathbf{J} \cdot \mathbf{I}\mu u_r, \dots,$$

lead to the following mass, momentum, and viscous volumetric densities:

$$\nabla \cdot \rho \mathbf{u} = \frac{1}{r}(r\rho u)_r + (\rho w)_z, \quad \nabla \cdot \rho \mathbf{u} \mathbf{u} = \frac{1}{r}(r\rho u u)_r + (\rho u w)_z$$

$$\text{and } \nabla \cdot \mu \nabla u = \frac{1}{r}(r\mu u)_r + (\mu u_z)_z, \dots \text{for } \mu = \mu(t, \mathbf{r}),$$

respectively.

Furthermore, in order to save records, as in [32], let us represent the matrices by double vectors, or more specifically by divectors: an identity matrix is the sum of the direct squares of the orts

$$\vec{\mathbf{e}} = \begin{pmatrix} 1 & 0 & 0 \\ 0 & 1 & 0 \\ 0 & 0 & 1 \end{pmatrix} = \mathbf{ii} + \mathbf{jj} + \mathbf{kk} = \mathbf{II} + \mathbf{JJ} + \mathbf{kk},$$

further, the matrix of momentum flow by the proportion $\rho\mathbf{uu}$ of $\mathbf{uu} = (u\mathbf{I} + v\mathbf{J} + w\mathbf{k})\mathbf{u}$ density of momentum flow

$$\nabla \cdot \rho \mathbf{uu} = \left(\nabla \cdot \rho u\mathbf{u} - \frac{\rho v^2}{r}\right)\mathbf{I} + \left(\nabla \cdot \rho v\mathbf{u} + \frac{\rho v u}{r}\right)\mathbf{J} + (\nabla \cdot \rho w\mathbf{u})\mathbf{k},$$

finally, the original for fluid mechanics the "matrix of the deformation rates" [33] by the matrix of the rate deformations, or fluid deformations $\mathbf{u_r}$ to be the sum of $\mathbf{u}_x\mathbf{i} + ..$ of direct products $\mathbf{u}_x\mathbf{i} + ..$:

$$\mathbf{u_r} = \begin{pmatrix} u_x^x & u_y^x & u_z^x \\ u_x^y & u_y^y & u_z^y \\ u_x^z & u_y^z & u_z^z \end{pmatrix} = \begin{pmatrix} u_x^x\mathbf{ii} + & u_y^x\mathbf{ij} + & u_z^x\mathbf{ik} + \\ u_x^y\mathbf{ji} + & u_y^y\mathbf{jj} + & u_z^y\mathbf{jk} + \\ u_x^z\mathbf{ki} + & u_y^z\mathbf{kj} + & u_z^z\mathbf{kk} \end{pmatrix} = \mathbf{u}_x\mathbf{i} + \mathbf{u}_y\mathbf{j} + \mathbf{u}_z\mathbf{k},$$

or

$$\mathbf{u_r} = \mathbf{u}_r\mathbf{I} + \frac{1}{r}\mathbf{u}_\theta\mathbf{J} + \mathbf{u}_z\mathbf{k},$$

with

$$\mathbf{u}_\theta = (u\mathbf{I} + v\mathbf{J} + w\mathbf{k})_\theta = u\mathbf{I}_\theta + v\mathbf{J}_\theta = u\mathbf{J} - v\mathbf{J} = \mathbf{k}\times\mathbf{u}$$
$$\text{and } \mathbf{u}_{\theta\theta} = \mathbf{k} \times \mathbf{k}_\theta = \mathbf{k} \times (\mathbf{k} \times \mathbf{k}) = \mathbf{k}\mathbf{k} \cdot \mathbf{u} - \mathbf{k} \cdot \mathbf{k}u = \mathbf{k}w - \mathbf{u},$$

with divergence

$$\nabla \cdot \mathbf{u}_r = \left(\frac{1}{r}(ru)_r + w_z\right)_r \mathbf{I} + \left(\frac{1}{r}(ru)_r + w_z\right)_z \mathbf{k} = (\nabla \cdot \mathbf{u})_r\mathbf{I} + (\nabla \cdot \mathbf{u})_z\mathbf{k} = \nabla\nabla \cdot \mathbf{u}$$

convolution or point product

$$\nabla\mu \cdot \mathbf{u}_r = (\nabla\mu \cdot \mathbf{u}_r)\mathbf{I} - \frac{\mu_r v}{r}\mathbf{J} + (\nabla\mu \cdot \mathbf{u}_z)\mathbf{k},$$

with divergence

$$\nabla \cdot \mu\mathbf{u_r} = \mu\nabla \cdot \mathbf{u_r} + \nabla\mu \cdot \mathbf{u_r} = \mu\nabla\nabla \cdot \mathbf{u} + \nabla\mu \cdot \mathbf{u_r}$$
$$= \left(\mu(\nabla \cdot \mathbf{u})_r + \nabla\mu \cdot \mathbf{u}_r\right)\mathbf{I} - \frac{\mu_r v}{r}\mathbf{J} + \left(\mu(\nabla \cdot \mathbf{u})_z + \nabla\mu \cdot \mathbf{u}_z\right)\mathbf{k},$$

and the conjugate matrix is the sum of the corresponding reverse multiplications

$$\nabla\mathbf{u} = \begin{pmatrix} u_x^x & u_x^y & u_x^z \\ u_y^x & u_y^y & u_y^z \\ u_z^x & u_z^y & u_z^z \end{pmatrix} = (\nabla\mu^x)\mathbf{i} + (\nabla\mu^y)\mathbf{j} + (\nabla\mu^z)\mathbf{k} = \mathbf{i}u_x + \mathbf{j}u_y + \mathbf{k}u_z,$$

or

$$\nabla\mathbf{u} = (\mathbf{u_r})_* = \mathbf{I}u_r + \frac{1}{r}\mathbf{J}u_\theta + \mathbf{k}u_z, \cdots \mathbf{u_r} = (\nabla\mathbf{u})_*,$$

with divergence

$$\nabla \cdot \mu\nabla\mathbf{u} = \left(\nabla \cdot \mu\nabla u - \frac{\mu u}{r^2}\right)\mathbf{I} + \left(\nabla \cdot \mu\nabla v - \frac{\mu u}{r^2}\right)\mathbf{J} + (\nabla \cdot \mu\nabla w)\mathbf{k}$$

so that

$$\nabla \cdot \mu(\mathbf{u_r} + \nabla\mathbf{u}) = \left(\begin{array}{c} \nabla \cdot \mu\nabla u - \frac{\mu u}{r^2} \\ + \mu(\nabla \cdot \mathbf{u})_r + \nabla\mu \cdot \mathbf{u}_r \end{array}\right)\mathbf{I} + \left(\begin{array}{c} \nabla \cdot \mu\nabla v \\ - \frac{(r\mu)_r v}{r^2} \end{array}\right)\mathbf{J} + \left(\begin{array}{c} \nabla \cdot \mu\nabla w \\ + \mu(\nabla \cdot \mathbf{u})_z \\ + \nabla\mu \cdot \mathbf{u}_z \end{array}\right)\mathbf{k}.$$

Let us consider the motion of the particles in the wet mixture. If we follow Euler and add the strength of the molecular pressure p to Newton's law of dynamics, then, taking into account conservation of mass or the continuity equation

$$\rho_t + \nabla \cdot \rho\mathbf{u} = 0$$

add the buoyant force of Archimedes $-\nabla p$ (or the force of molecular pressure p) to the body force (or, to be exact, $\mathbf{g}\rho dV$) applied to mass ρdV with a given acceleration $\mathbf{g} = \mathbf{g}(t, \mathbf{r})$ in the Newton's dynamic law that balances the inertial force $\rho\mathbf{a}$ of fluid acceleration $\mathbf{a} = \mathbf{u}_t + \mathbf{u} \cdot \nabla\mathbf{u}$ with the sum of $\rho\mathbf{g}$ and $-\nabla p$ refined subsequently by both the force $\nabla \cdot \mu\vec{\mathbf{b}}$ of the Newton's dynamic viscosity $\mu = \mu(t, \mathbf{r})$ (supposed to be constant initially) and the matrix

$$\vec{\mathbf{b}} = \mathbf{u_r} + \nabla\mathbf{u} - \frac{2}{3}(\nabla \cdot \mathbf{u})\vec{\mathbf{e}},$$

of Navier ($\nabla \cdot \mathbf{u} = 0$) and Stokes and the relevant pressure force $-\nabla(p - \varsigma\mu\nabla \cdot \mathbf{u})$ with the second, or volume viscosity $\varsigma\mu$, $\varsigma = \text{cons} \geq 0$ [26–28].

As a result, with appropriate contact (or applied to coordinate areas) forces to be tensions, stresses, or components of the Euler–Navier–Stokes matrix

$$\vec{\mathbf{p}} = (p - \varsigma\mu\nabla \cdot \mathbf{u})\vec{\mathbf{e}} - \mu\vec{\mathbf{b}} = P\vec{\mathbf{e}} - \mu(\mathbf{u_r} + \nabla\mathbf{u}),\ P = p + \left(\frac{2}{3} - \varsigma\right)\mu\nabla \cdot \mathbf{u},$$

with divergence

$$\nabla \cdot \vec{\mathbf{p}} = \nabla \cdot P\vec{\mathbf{e}} - \nabla \cdot \mu(\mathbf{u_r} + \nabla\mathbf{u}) = \nabla P - \nabla \cdot \mu(\mathbf{u_r} + \nabla\mathbf{u}),$$

and ensures the dynamic equilibrium of a continuous medium in the corresponding Navier–Stokes equations:

$$(\rho\mathbf{u})_t + \nabla \cdot (\rho\mathbf{uu} + \vec{\mathbf{p}}) = \rho\mathbf{g} \text{ and } \rho_t + \nabla \cdot \rho\mathbf{u} = 0, \qquad (2.1)$$

or

$$(\rho u)_t + \left(\tfrac{1}{r}(r\rho uu)_r + (\rho uw)_z - \tfrac{\rho v^2}{r}\right) + P_r$$
$$- \left(\tfrac{1}{r}(r\mu u_r)_r + (\mu u_z)_z - \tfrac{\mu u}{r^2} + \mu\left(\tfrac{1}{r}(r\mu)_r + (w)_z\right)_r + \mu_r u_r + \mu_z w_r\right) = \rho g^r,$$
$$(\rho v)_t + \left(\tfrac{1}{r}(r\rho vu)_r + (\rho vw)_z + \tfrac{\rho uv}{r}\right) - \left(\tfrac{1}{r}(r\mu v_r)_r + (\mu v_z)_z - \tfrac{(r\mu)_r v}{r^2}\right) = 0,$$
$$(\rho w)_t + \left(\tfrac{1}{r}(r\rho wu)_r + (\rho ww)_z\right) + P_z$$
$$- \left(\tfrac{1}{r}(r\mu w_r)_r + (\mu w_z)_z + \mu\left(\tfrac{1}{r}(ru)_r + w_z\right)_z + \mu_r u_z + \mu_z w_z\right) = \rho g^z,$$
$$P = p + \mu\left(\tfrac{2}{3} - \varsigma\right)\left(\tfrac{1}{r}(ru)_r + w_z\right) \text{ and } \rho_t + \tfrac{1}{r}(r\rho u)_r + (\rho w)_z = 0,$$

Contact stresses of matrix **P** are naturally attributed to the source of liquid strains $\mathbf{u_r}$ for particles of the wet mixture in dynamic equilibrium (Eq. 2.1), which replaces static equilibrium (Hooke's law) for small displacements in the solid particles of the dry material [33].

The distortion of the velocity field assessed using measurement of liquid deformation [32],

$$D = D[\mathbf{u}] = \frac{1}{2}\|\mathbf{u_r} + \nabla\mathbf{u}\|^2 = \frac{1}{2}|\mathbf{u}_x + \nabla u^x|^2 + \frac{1}{2}|\mathbf{u}_y + \nabla u^y|^2 + \frac{1}{2}|\mathbf{u}_z + \nabla u^z|^2, \ldots$$

which is a familiar function measuring the dissipation of mechanical energy into thermal energy [27],

$$D = 2\left(u_x^x\right)^2 + 2\left(u_y^y\right)^2 + 2\left(u_z^z\right)^2 + \left(u_x^y + u_y^x\right)^2 + \left(u_y^z + u_z^y\right)^2 + \left(u_z^x + u_x^z\right)^2$$

or, in the axisymmetric polar variables

$$D = 2u_r^2 + 2\left(\frac{u}{r}\right)^2 + 2w_z^2 + \left(v^r - \frac{v}{r}\right)^2 + (v_z)^2 + (u_z + w_r)^2,$$

and systematically used, as shown, for example, in [34], the argument (or independent variable) of viscous rheology, suggesting the existence of an equation to measure the plastic state of the material [28–30]:

$$\mu = \mu(\gamma), \ \gamma = D[\mathbf{u}] \geq 0.$$

The geometric structure of the argument $\gamma = D$ reveals the identity of measure, as directly verifiable resolving D the sum of the proportions [32],

$$D = \frac{2}{3}A^2 + B^2 + \frac{2}{3}C^2$$

squares of local norms of inhomogeneity

$$A = \sqrt{\left(u_x^x - u_y^y\right)^2 + \left(u_y^y - u_z^z\right)^2 + \left(u_z^z - u_x^x\right)^2},$$

shear

$$B = \sqrt{\left(u_x^y - u_y^x\right)^2 + \left(u_y^z - u_z^y\right)^2 + \left(u_z^x - u_x^z\right)^2},$$

and compressibility

$$C = u_x^x + u_y^y + u_z^z$$

for liquid deformations, described by the matrix $\mathbf{u_r}$.

It is possible to show [15] that with the help of proper deformations of inhomogeneity, shear, and compressibility, material mixed in the arbitrary volume V,
produces work in a second, or power of deformations

$$W = \int_V \mu \left(\frac{2}{3} A^2 + B^2 + \varsigma C^2 \right) dV.$$

Let us simplify the initial problem. As already noted, in addition to transportation,
the main purpose of the extruder is to compact the material in zone 2 (Fig. 2.20).
We will limit ourselves to the portion (length) of zones 3 and 4, suggesting constant
density of material, approximated by Newtonian rheology:

$$\rho = \text{const} > 0 \, (\text{therefore, } \nabla \cdot \mathbf{u} = 0) \text{ and } \mu = \text{const} > 0.$$

We consider the flow of material in the space between two concentric infinite
cylinders. Further, with the stationary external cylinder $r = a = \text{const} > 0$, we will
maximally the auger operation for the transportation and shearing of mass in the
extruder or between cylinders

$$\varepsilon a < r < a, \; \varepsilon, a = \text{const} > 0, \; \varepsilon < 1,$$

shifting it to the inner cylinder $r = \varepsilon a$, that, due to conditions of liquid particles
sticking to it, now moves this mass along the axis z, at the same time as rotating it
around this axis with a constant velocity of boundary displacement $w_- > 0$ and
torsion $v_- < 0$, as in Fig. 2.23:

$$w|_{r=\varepsilon a} = w_-, w|_{r=a} = w_+, v|_{r=\varepsilon a} = v_-, \quad w_\mp, v_\mp = \text{const},$$
$$\text{where } w_- > 0, \quad v_- < 0 \text{ and } w_+ = v_+ = 0. \tag{2.2}$$

Finally, assume that the incoming mass is pushed further with a constant gradient of
excess pressure, or by the difference of its values $p_\pm$ at the ends of the chosen
length:

$$-p_z = \frac{p_+ - p_-}{l} = \text{const} > 0, 0 < z < l, p_- = p|_{z=0}, p_+ = p|_{z=l}. \tag{2.3}$$

In the absence of mass forces and radial displacements of the liquid material
rotating and being moved longitudinally by the auger, and more precisely, under the
assumption that

$$u = 0, \; v = v(r), \; w = w(r) \text{ and } g^r = g^z = 0, \tag{2.4}$$

the dynamic equilibrium (2.1) with conditions (2.2)–(2.4) is reduced to the
proportions:

$$-\frac{1}{r}(rw_r)_r = \frac{-p_z}{\mu}, \ (rv_r)_r = \frac{v}{r}, \ p_r = \frac{\rho v^2}{r}, \ -p_z = \frac{p_+ - p_-}{l}, \quad \varepsilon a < r < a,$$

$$0 < z < l, \ w(\varepsilon a) = w_-, w(a) = w_+ = 0, \ v(\varepsilon a) = v_-, \ v(a) = v_+ = 0,$$

or to Hagen–Poiseuille's flow between the pipes [26, 35],

$$w(r) = w_+ - \frac{p_z(a^2 - r^2)}{4\mu} + \left(w_- - w_+ + \frac{p_z a^2(1 - \varepsilon^2)}{4\mu}\right)\frac{\ln\frac{r}{a}}{\ln\varepsilon} \text{ with } w_+ = 0,$$

at the same time spiral, as the Couette flow with azimuthal velocity

$$v(r) = \frac{\varepsilon a}{r}\frac{v_- - \varepsilon v_+}{1 - \varepsilon^2} + \frac{r}{a}\frac{v_+ - \varepsilon v_-}{1 - \varepsilon^2} = \frac{\varepsilon v_-}{1 - \varepsilon^2}\left(\frac{a}{r} - \frac{r}{a}\right) \text{ with } v_+ = 0,$$

and with pressure falling along the radius to the center of rotation, as in a centrifuge [31]

$$p(r,z) = p_+ - \frac{z}{l}(p_+ - p_-) - \int_r^a \frac{\rho v^2(r')}{r'}dr', l \quad p_r = \frac{\rho v^2}{r} > 0,$$

$$\varepsilon a < r < a, 0 < z < l.$$

In Fig. 2.24 in dimensionless variables of the radius X and speeds of displacement and rotation of auger, Z and Y,

$$\frac{4\mu w}{-p_z z^2} = Z = 1 - X^2 - \frac{1 - \varepsilon^2 - \delta}{\ln\varepsilon}\ln X, \quad 0 < \delta = \frac{4\mu w}{-p_z a^2} < 1 - \varepsilon^2,$$

and

$$\frac{v(r)}{v_-} = Y = \frac{\varepsilon}{1 - \varepsilon^2}\left(\frac{1}{X} - X\right)$$

given the dependence of the latter from the first, i.e., dimensionless profiles $Z(X)$ and $Z(X)$ of the azimuthal $w(r)$ and axial $v(r)$ components of the velocity $\mathbf{u}$. The latter depends on dimensionless parameters ε and δ: ratio of the inner radius r_{min} to an external a, or radial factor

$$\varepsilon = \frac{r_{min}}{a}, 0 < \varepsilon < 1,$$

and the ration of velocity of the boundary displacement w_- to the following speed viscous pressure w_p, or a dynamic factor

$$\delta = \frac{w_-}{w_p} = \frac{4\mu w_-}{-p_z a^2} > 0 \text{ at } w_p = \frac{-p_z a^2}{4\mu} \text{ and } -p_z > 0.$$

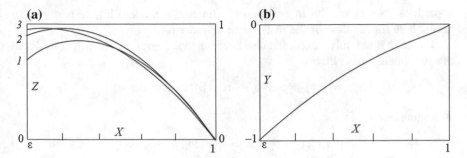

Fig. 2.24 Profiles of axial (left) and azimuthal (right) component of the considered spiral flow at two subcritical (1 and 2) and one supercritical (3) values of dynamic factor

Considered SVE extruders [30] are operated with the usually moderate values of $\varepsilon \approx 0.5$ and mostly at the subcritical value of $0 > \delta < 1 - \varepsilon^2$, rarely, at the supercritical factor $\delta > 1 - \varepsilon^2$, that is illustrated by the dimensionless dependence $Z(X)$ in Fig. 2.24, left.

At the subcritical dynamic factor and the maximum possible speed of material transportation by the auger $w(r)$ is limited by value

$$w_* = \max_{\varepsilon a \leq r \leq a} w(r) = w_p \max_{\varepsilon \leq X \leq 1} Z(X) = w(r_*),$$

close to the speed of viscous flow w_p, achieved within the field of flow: $\varepsilon a < r_* < a$. With the growth of the factor $\delta < 1 - \varepsilon^2$ this value approaches the speed w_p, shifting simultaneously on the internal radius of the cylinder $r = \varepsilon a$ and reaches w_p on the border of flow area, i.e., $r_* = \varepsilon a$ at $\delta = 1 - \varepsilon^2$, remaining maximum speed of the transported mass $w_* = w_p$ and at any superficial factor $\delta \geq 1 - \varepsilon^2$.

Thus, increasing the speed of displacement w_- at superficial factor δ does not increase the speed of transportation $w(r)$ in the flow area $\varepsilon a < r - a$, reaching the maximum value w_* on the border of this area $r_* = \varepsilon a$ and equal to the speed of viscous flow w_p.

The above concepts came from the central assumptions of continuum mechanics about the presence of the molecular pressure p in such materials and produced by its flow $\mathbf{u}$ liquid deformations $\mathbf{u_r} = (\nabla \mathbf{u})_*$, which together generate contact strain $\vec{\mathbf{p}}$, balancing the resulting acceleration of the particles in the Navier–Stokes Equations (2.1). The original premise of the mechanics of viscoplastic deformation seems to be the opposite: there is a solid body and the tension applied to its surface is defined as $\vec{\mathbf{p}}$ (Shear stress), which produce deformation $\mathbf{u_r}$ (Shear rate) upon reaching of a critical level—the so-called plasticity limit $\|\vec{\mathbf{p}}\| = const$ (Yield stress). But details of the dependence between tension and deformation are reduced to the above classical formula for $\vec{\mathbf{p}}$. The same viscoplastic state of the material is described by the above-mentioned dependence of dynamic viscosity $\mu = \mu(\gamma)$ on measure of deformations $\gamma = D[\mathbf{u}]$, reduced to the constant $\mu = const$ at the superficial level of tension in well-known Bingham's model [28], which we also accept here for the flow of the material in extruder [36].

The mass is not only mixed, but also compressed. Therefore, we will consider in the simplified variant that

$$\rho = \rho(r) \neq \text{const} \, (\nabla \cdot \mathbf{u} \neq 0) \text{ and } \mu = \text{const}.$$

Assuming that

$$u = u(r), \upsilon = \upsilon(r), w = w(r) \text{ and } g^r = g^z = 0,$$

then the continuity equation in (2.1) leads to a permanent

$$r\rho u = m = \text{const}, \quad or \, ru = m\varphi, \text{ where } \varphi = \frac{1}{\rho}$$

specific volume. When divided by dynamic viscosity,

$$\alpha = \frac{m}{\mu},$$

is a dimensionless value. Having in (2.1)

$$p_r = \frac{\rho \upsilon^2}{r} \text{ and } \frac{\alpha + 1}{r^2} (ru)_r - \left(\frac{1}{3} + \varsigma\right) \left(\frac{1}{r} (ru)_r\right)_r - \frac{1}{r}(ru)_{rr} - \frac{\alpha ru}{r^3} = 0$$

we solved for the values $\varphi_\mp$:

$$\varphi = \frac{\varepsilon^{\lambda+} \varphi_+ - \varphi_-}{\varepsilon^{\lambda+} - \varepsilon^{\lambda-}} \left(\frac{r}{a}\right)^{\lambda-} + \frac{\varepsilon^{\lambda-} \varphi_+ - \varphi_-}{\varepsilon^{\lambda-} - \varepsilon^{\lambda+}} \left(\frac{r}{a}\right)^{\lambda+},$$

$$\lambda_\mp = \frac{8 + 6\varsigma + 3\alpha}{2(4 + 3\varsigma)} \mp \sqrt{\left(\frac{8 + 6\varsigma + 3\alpha}{2(4 + 3\varsigma)}\right)^2 - \frac{3\alpha}{4 + 3\varsigma}} at \left(\frac{8 + 6\varsigma + 3\alpha}{2(4 + 3\varsigma)}\right)^2 > \frac{3\alpha}{4 + 3\varsigma},$$

Satisfying the conditions (2.2)–(2.3) and Eq. (2.1)

$$\frac{m}{r}\upsilon_r + \frac{m\upsilon}{r^2} - \frac{\mu}{r}(r\upsilon_r)_r + \frac{\mu\upsilon}{r^2} = 0, \quad \frac{m}{r}w_r + \frac{mw}{r^2} - \frac{\mu}{r}(rw_r)_r = -P_z = -p_z,$$
$$P = p + \mu\left(\frac{2}{3} - \varsigma\right)\frac{1}{r}(ru)_r, \, \varepsilon a < r < a,$$

axial and azimuthal velocity components, in this case, have profiles:

$$w = \frac{\varepsilon^{\gamma+} w_+ - w_-}{\varepsilon^{\gamma+} = \varepsilon^{\gamma-}} \left(\frac{r}{a}\right)^{\gamma-} + \frac{\varepsilon^{\gamma-} w_+ - w_-}{\varepsilon^{\gamma-} - \varepsilon^{\gamma+}} \left(\frac{r}{a}\right)^{\gamma+}$$
$$+ \frac{-p_z a^2}{\mu(4 - \alpha)} \left(\frac{\varepsilon^{\gamma+} - \varepsilon^2}{\varepsilon^{\gamma+} - \varepsilon^{\gamma-}} \left(\frac{r}{a}\right)^{\gamma-} + \left(\frac{\varepsilon^{\gamma-} - \varepsilon^2}{\varepsilon^{\gamma-} - \varepsilon^{\gamma+}}\right) \left(\frac{r}{a}\right)^{\gamma+} - \left(\frac{r}{a}\right)^2\right),$$

$$\gamma_\mp = \frac{\alpha}{2} \mp \sqrt{\left(\frac{\alpha}{4} - 1\right)\alpha} \, at \left(\frac{\alpha}{4} - 1\right)\alpha > 0,$$

Fig. 2.25 Raw brex exiting
the extruder

and

$$v = \frac{\varepsilon^{\beta_+}v_+ - v_-}{\varepsilon^{\beta_+} - \varepsilon^{\beta_-}}\left(\frac{r}{a}\right)^{\beta_-} + \frac{\varepsilon^{\beta_-}v_+ - v_-}{\varepsilon^{\beta_-} - \varepsilon^{\beta_+}}\left(\frac{r}{a}\right)^{\beta_+},$$

$$\beta_{\mp} = \frac{\alpha}{4} \mp \left(\frac{\alpha}{2} + 1\right) = -\frac{\alpha}{4} - 1, \frac{3\alpha}{4} + 1.$$

The received analytical dependences allow one to qualitatively assess the flow of
the formable mass in the extruder and can be used for further numerical simulation.

In zone 5, the brex are squeezed out of the holes in the die (Fig. 2.25), which
completes the process of their formation.

2.4   The Length and the Cross-sectional Shape of Brex

The considerable length of the brex at the die exit is notable. In work [9], we carried
out a finite-element simulation of exit of elongated brex from an extruder using the
SIMULIA Abaqus software complex [37]. This modeling allowed us to offer a
satisfactory description of the mechanism by which an elongated brex breaks into
several brex of shorter length. When bending an elongated brex under the force of
its own weight, zones of maximum stress and sources of further breakage are
formed in its body. Figure 2.26 shows the results of mathematical modeling of the
various stages of product breakage, both at the exit from the extruder and upon
falling on the supporting surface (floor, conveyor belt, etc.).

In extrusion agglomeration, it is possible to produce brex of practically any
shape by means of simply changing the shape of the orifice through which it passes
in the die.

Fig. 2.26 Simulation of pressures in an elongated brex (on the left), its breakdown at the exit from the extruder (in the middle), and its falling onto a supporting surface (on the right)

Earlier, we investigated the effects of the shape of the brex on its strength [8]. Our experiment demonstrated that there was a significant difference in the strength between brex from the same batch depending on whether it has round or oval shapes. The strength of oval-shaped brex is dependent on the direction in which pressure is applied to it. When applying a crushing force along the minor axis of the oval, it may take almost twice as much pressure to achieve the same stress levels as it does when applying a pressure along the major axis. It can be assumed that the difference in the values of the strength can be attributed to the difference in the area of contact surface during crushing. For circular cross sections, the contact patch has a smaller area than it does for the oval samples if applied along the minor axis. Thus, when the load is equally distributed along the minor axis, the acting stresses in the oval-shaped brex are smaller than they are in the round shape. It is also clear that when the crushing force is directed along the major axis of the oval, the magnitude of stresses will be correspondingly higher. We performed a finite-element simulation of a full-scale brex-crushing experiment using the SIMULIA Abaqus software. The objects of our focus were brex of the same area with round and oval shapes. The ratio of the length of the major and minor axes of the oval is 2:1. Finite-element modeling was performed using a linear-elastic material model. As a result of mathematical modeling, we were able to evaluate the distribution of the equivalent von Mises stress over the cross section of the briquettes (Fig. 2.27). If the pressure required to split a round brex was 1, then the relative splitting pressures for an oval brex were 1.33 and 0.67 respectively along the minor and major axes.

Fig. 2.27 Possible stresses on oval brex (left and middle) and circular cross sections, and the distribution of equivalent stresses in the body of the brex

2.5 Migration of Fines

Another feature of SVE is the possibility of extrusion of the fine fraction of the mixture during the production of brex on its surface, as illustrated by photos of brex made from a mixture of carbon-based material and construction fines (Fig. 2.28, left). It is evident that the carbon-based portion of the mixture formed the outer layer of the brex. The image on the right depicts coke brex bound by epoxy resin, the smallest particles of which are pushed out during the process of extrusion.

To understand the process that occurs when a mixture of large and small particles is moved, we proposed a simplified hydrodynamic model to describe the motion of large particles in a liquid. A similar model was developed for separating particles in dust-cleaning systems [38]. When applying the hydrodynamic model, we allowed for a simplification in which the fine phase is replaced by a liquid. In reality, the behavior of a fine solid phase differs from the behavior of a liquid, mainly because the solid phase is difficult to deform and does not flow as easily as a liquid, especially around obstacles. For a complete assessment, it is necessary to take into account the effect of particle shape and a number of other factors. However, for a general understanding of the principles of particle motion and of the forces that affect particles, such a simplification is permissible.

When the molded mass passes through the opening of die, we see a flow profile resembling that of the Poiseuille law on the flow of water through pipes (Fig. 2.29). A characteristic feature of such profiles is a large velocity gradient near the wall and a zero gradient at the center of the tube.

When finite-dimensional particles move in a gradient flow in an ordinary (Newtonian) environment, the resulting lift force is similar to that which acts on the wing of an aircraft. The latter occurs due to the difference in pressure under and above the wing of the aircraft ([26] and Fig. 2.30).

Fig. 2.28 Brex with a layer of fine fraction on the surface. Left: coal (surface layer) and fine fraction of construction materials. Right: coke brex with epoxy resin as binder

Fig. 2.29 Poiseuille flow profile

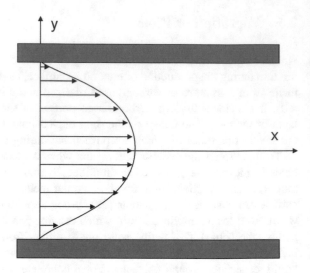

Fig. 2.30 Flow along the plate [26]

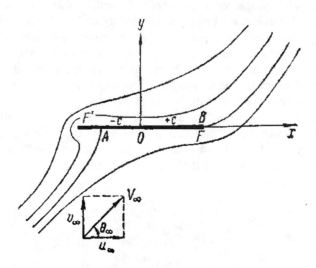

As per Bernoulli's principle, an ideal fluid of zero viscosity maintains pressure equal to the sum of its hydrostatic and dynamic pressures (the so-called Bernoulli integral) along its streamline [26].

$$p + \frac{\rho u^2}{2} = \text{const.} \tag{2.5}$$

Since air moves more slowly beneath the wing, the hydrostatic pressure must be greater to keep the Bernoulli integral (5) constant, with the reverse being true of the environment above the wing. For a particle in a flow with a velocity gradient (Fig. 2.31), an analogous situation arises: the pressure below the particle will be

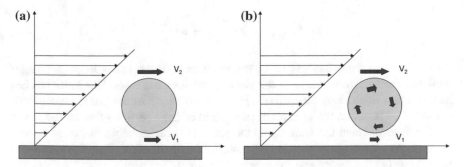

Fig. 2.31 **a** Flow around the particle near the wall extending from the hole; **b** The particle rotated by the flow

greater than it is above it. In this case, the particle may be pulled into rotation as a result of the velocity gradient. In this case, we will see a secondary lift force in the form of the Magnus effect, which surrounds a rotating particle (Fig. 2.31b).

As per Joukowski's theorem, lift is determined by the flow density ρ, circulation of velocity around the particle J, and the velocity of the oncoming flow u (the stream is assumed to have a horizontal velocity of u and a vertical velocity of $u = 0$) [26]:

$$F_{Magnus} = \rho J u \qquad (2.6)$$

The circulation around the particle as it rotates in the flow is given by:

$$J = \frac{\pi r^2}{2} r \frac{du}{dy} \qquad (2.7)$$

where

πr^2 is a mid-length section of the particle,
r is the linear dimension of the particle,
du/dy is the velocity gradient of the flow (the y axis is directed vertically).

As a result, the Magnus force is proportional to the cube of the particle's volume, the velocity of the oncoming flow, and the velocity gradient [39]

$$F_{Magnus} \sim \rho \frac{\pi r^3}{2} u \frac{du}{dy} \qquad (2.8)$$

In reality, it is necessary to take into account that the particle is carried by the flow and has a horizontal velocity U (i.e., a relative velocity $u–U$). Therefore, we ultimately have the following:

$$F_{\text{Magnus}} = \rho \frac{\pi r^3}{2} (u - U) \frac{du}{dy} \qquad (2.9)$$

As can be seen from formula (2.9), the magnitude of the lift is larger for larger particles. Force also increases as the velocity gradient increases. The latter reaches its maximum value along the wall of a pipe, while amounting to zero at the center of the stream. Therefore, the greatest displacement of large particles toward the center will be observed near the wall, while the smallest particles being displaced toward the wall. As a result, there should be a nonuniform distribution of particles, with the largest ones at the center and the smallest ones near the wall.

When the particles move slowly relative to the "liquid" (in our case, relative to the fine phase), the Stokes' frictional force is observed [26]:

$$F_{\text{Stokes},x} = 6\pi\mu r(u - U), \qquad (2.10)$$

where μ is dynamic viscosity (analogous to the friction of a small particle against a large particle), $(u-U)$ horizontal velocity relative to the flow.

Similarly, in the vertical direction, Stoke's frictional force is measured as:

$$F_{\text{Stokes},x} = 6\pi\mu r V, \qquad (2.11)$$

where V is the vertical flow velocity relative to the object.

From the equilibrium of the Magnus force and Stokes's frictional force (where $F_{\text{Magnus}} = F_{\text{Stokes},y}$, y), we derive a formula for estimating velocity of the particle's transverse (vertical) displacement:

$$V = \frac{\rho}{12\mu} (u - U) \frac{du}{dy} r^2 \qquad (2.12)$$

This formula maintains that displacement velocity is proportional to the square of the particle size

Now we can make a qualitative assessment of the relative displacements of large and small particles. We can say that in (2.12) the difference $(u-U)$ characterizes the velocity of small particles relative to large particles. The velocity gradient du/dy is the same for small and large particles. Let us take, for example, the particles "large" and "small", where "large" is three times are large as "small" ($r_{\text{large}} = 3r_{\text{small}}$). Since $V \approx r^2$ and all other factors are equal, the transverse velocities of large and small particles will differ by a factor of nine.

$$V_{\text{large}} = -\frac{r_{\text{large}}^2}{r_{\text{small}}^2} V_{\text{small}} = -9V_{\text{small}}. \qquad (2.13)$$

The "minus" sign is necessitated by the multidirectional motion of large and small particles in accordance with formula (2.12). In reality, we can assume that small particles remain in place and large ones are squeezed into the center of the stream.

The velocities of the relative transverse displacement of the particles are proportional to the square of the ratio of the particles' volumes.

The phenomenon of small fraction displacement is explained by the proposed hydrodynamic model using particles of two different sizes.

2.6 Transportation, Warehousing, and Storage of Brex

The high density of freshly made (green) brex enables them to be delivered immediately to the storage site by conveyor or loaded onto dump trucks (Fig. 2.32) for transport to the warehouse.

Due to the viscous or plastic reaction of a mass of brex to external mechanical action described in Sect. 3.1.—a reaction attributed to the granulometric composition of the mixture and the presence of bentonite—the brex show a sufficiently high impact strength. In the technological cycle of production of the brex for the smelting of ferromanganese (12.4% of aspiration dust from the production of ferromanganese, 40.8%—manganese containing sludge, 43.4%—screening of manganese pellets, and 3.4% bentonite as a binder) can withstand dropping from a height of 5 m onto the concrete floor of the bunker (Fig. 2.33).

Using the SIMULIA Abaqus software complex, it is possible to simulate the process of unloading oval-shaped brex in order to calculate the statistics surrounding the placement of brex on a pile (Fig. 2.34) [9]. Our results revealed that 56.25% of brex have a "flat" orientation (in which the long axis of an oval lies horizontally), 12.5% are oriented with the long axis perpendicular to the base, 12.5% are on the bottom, and 18.75% on the "edge." Thus, we were able to show that fewer than 13% of the brex in any given stack would lie in the direction in

Fig. 2.32 Loading of freshly formed brex onto a dump truck and its subsequent unloading 5 min later

Fig. 2.33 Unloading of green brex in the bunker with a height of over 5 m

Fig. 2.34 Simulation
modeling of placement of
oval brex on the bed of the
furnace

which the crushing force would theoretically be directed along the oval's long axis.
We also noted that when an oval shape is formed, the free surface area increases
(for brex with a 2:1 axis ratio, the surface area growth amounts to 5.8%), which
would positively affect the process of its reduction in the reducing atmosphere.

In accordance with Russian standard 2787 (GOST 2787) [40], brex are dropped
three times (freefall) from a height of 1.5 m onto a metal or concrete slab to
determine their frailness. They should not generate more than 10% (mass) of fines.
From every five brex dropped, four must pass the test. To simulate the process of
brex breakage under impact loading, we evaluated the fall of brex onto an abso-
lutely rigid flat surface from a height of 1.5 m. The specimens were then assessed
from numerous angles: from the lateral surface, the front, and the edge (Fig. 2.35).

Fig. 2.35 Models of possible manners in which brex may fall onto a surface

The simulation was carried out using the elastic or plastic model for the brex behavior and keeping in mind the possibility of its destruction in the SIMULIA Abaqus software [41]. The physico-mechanical parameters of the material were determined by way of experimental data of the brex samples with a composition of 15% chromium ore concentrate, 15% coal, 67% aspiration dust resulting from ferrochrome production, and 3% Portland cement. In the experiment, the following properties of materials were noted using a bench-type single-column electrome-chanical Instron 3345 (USA) test machine with a loading capacity of 5 kN: a Young's modulus of 123 MPa, a Poisson's ratio of 0.3, and an ultimate strength 2.1 MPa. The results of the simulation are related in Figs. 2.36, 2.37 and 2.38. Results of simulations in which the brex fell from a height of 0.5 m and 2.0 m were also provided for the sake of comparing the damage parameter. The damage parameter is a conventional value characterizing the degree of material damage (0 for intact, 1 for completely destroyed). In the above simulation, regions of irreversible (inelastic) deformation were observed in contract zones in the body of the brex. A fall from a height of 0.5 m resulted in the complete destruction of the edge

Fig. 2.36 The distribution of damage in a cylindrical-shaped brex. Type of impact: lateral surface. Height of the fall: 0.5 m (left), 1.5 m (middle), and 2.0 m (right)

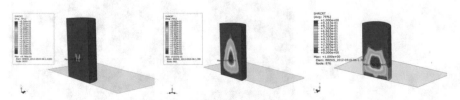

Fig. 2.37 The distribution of damage in a cylindrical-shaped brex. Type of impact: front. Height of the fall: 0.5 m (left), 1.5 m (middle), and 2.0 m (right)

Fig. 2.38 The distribution of damage in a cylindrical-shaped brex. Type of impact: edge. Height of the fall: 0.5 m (left), 1.5 m (middle), and 2.0 m (right)

of the brex only if it were to fall onto its front. This is due to the small area of contact and a resulting higher degree of mechanical stresses in the model. In the remaining cases, the fall of brex led to an irreversible degradation of its strength in the contact zone without initiating the elements' progressive destruction. At a drop from a height of 1.5 m, the destruction of the brex's thin superficial layers was observed in practically all cases. It is worth noting that when brex falls onto its front (Fig. 2.37), damage (without destruction) is localized at its center. This is explained by the interference of compression-extension waves upon impact. The application of mathematical modeling to assess the destruction of brex under impact loading gives us an idea of the frailness of brex and allows us to predict its strength upon impact.

Figure 2.39 shows the results of sieving portions of brex No. 2 (Table 2.7) after the brex was dropped four times from a height of 1.5 m. The test produces an insignificant amount (less than 2%) of fines of less than 5 mm, as well as a single fragment exceeding 12.5 mm.

In accordance with Russian standard 25471-82 (GOST 25471-82) [42], in order to determine the durability of iron ores, agglomerates, and pellets upon impact, the materials must be subjected to a test of triple falls from a height of 2 m. A test of ore-sludge brex (consisting of 27% iron ore concentrate, 34% sludge, 32% dust, 6% Portland cement, and 1% bentonite) showed that fine formation did not exceed 2.7% after 48 h in the absence of a preliminary homogenization of the moldable mixture. In the presence of a preliminary homogenization, fine formation was at

Fig. 2.39 Test results for a four-time drop from a height of 1.5 m of brex No. 2

Fig. 2.40 Breaking of brex (after 48 h of preparation) after a five drops from a height of 2 m

2.1%. When the material was dropped five times, these values were at 5.6% for non-homogenized brex mixture and 3.7% for homogenized brex mixture (Fig. 2.40). These results were obtained 48 h after the brex were manufacture. It is known that the strengthening of the cement binders bonded briquettes is achieved upon a 16–20 h heat treatment or by means of natural strengthening (at a temperature of no less than 20 °C) for a period of 7 days [43]. An important feature of the SVE method is that, during the first 4 days of hardening, the brex show a viscous-plastic character during tests, which is evidenced by their high impact strength. Furthermore, as mentioned previously in this Chapter, within 48 h, cement-bentonite binder brex achieve a maximal local compressive strength.

One of the shortcomings of SVE that manifests itself in cold conditions is the need to maintain the temperature of the building in which the extrusion equipment is located at least 5 °C. Otherwise, the very extrusion of the wet mass will be difficult if not impossible, and, in the event that the mixture contains Portland cement, the hardening of the brex will slow down or cease entirely.

In order to estimate the temperature in a pile of freshly made brex over a 2-day period of storage, we sought to measure nonstationary heat conduction in the natural convective heat transfer with the environment, taking into account the physical heating of the brex during extrusion and ignoring the heat released during cement hydration.

Our variable was ambient temperature at 0, 5, and 10 °C. The initial temperature of the brex at the time of storage is 25 °C.

The thermophysical properties used are shown in Table 2.9.

The evaluation of the heat fields in the pile of brex was carried out over the course of 5 days (432,000 s) under unchanged external conditions. In accordance with the technological conditions of the SVE line operation for blast furnace unit, the brex are taken to charge after two days of storage.

Table 2.9 Thermophysical properties of materials

Material	Thermal conductivity, W/(m K)	Heat capacity, J/kg K	Density, k/m^3
Brex	10	620	2500
Air	0.026	1005	1.18

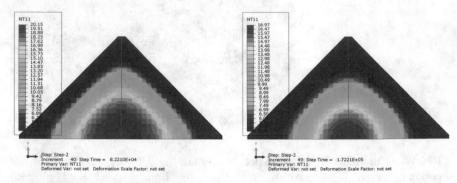

Fig. 2.41 Temperature field distribution. Day one is on the left-hand side; day two is on the right-hand side. The room temperature is 0 °C

Figure 2.41 shows the temperature distribution in the stack of brex at a room temperature of 0 °C after 24 and 48 h of storage.

Figure 2.42 shows the temperature distribution in the stack of brex at a room temperature of 5 °C after 24 and 48 h of storage.

Figure 2.43 shows the temperature distribution in the stack of brex at a room temperature of 10 °C after 24 and 48 h of storage.

Starting from a room temperature of 5 °C, the physical heat of the brex in the stack helps maintain sufficient temperatures for the hydration of cement. To verify the validity of this conclusion, we tested the durability of brex kept at 5 °C in a refrigeration chamber for 48 h by dropping them from a height of 2 m. The composition of this brex was as follows: 62.4% ore-sludge mixture, 30.6% blend of

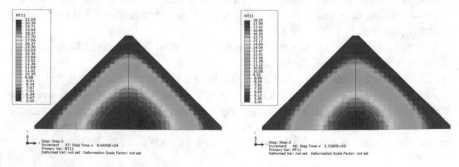

Fig. 2.42 Temperature field distribution. Day one is on the left-hand side; day two is on the right-hand side. The room temperature is 5 °C

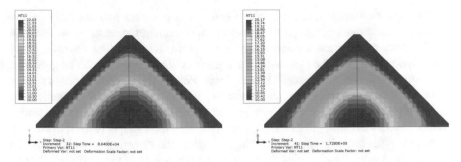

Fig. 2.43 Temperature field distribution. Day one is on the left-hand side; day two is on the right-hand side. The room temperature is 10 °C

blast furnace dust and dust resulting from aspiration, 6% Portland cement, and 1% bentonite. The amount of fines (particle less than 5 mm in size) formed after a three drop this the specified brex was 7% after 24 h of preparation at 5 °C, 4.3% after 48 h of preparation at 5 °C, and 4.2% after 72 h of preparation at 5 °C. If the brex are taken for charging into the blast furnace after 48 h of hardening at a room temperature of 5 °C, the strength of the brex is an acceptable level for use as blast furnace briquettes.

At temperatures close to zero, a significant number of brex will cool during the first 48 h to the temperatures at which hydration slows down or ceases entirely.

2.7 Conclusions

1. Allowing the briquetted material to develop plasticity is one of the most important aspects of preparing it for the production of brex. In many cases, it is sufficient to simply further grind the material. The method of grinding material for briquetting can affect the size, shape, and topography of the particle's surface. The degree to which these traits are affected depends on the features of the material (first and foremost on its porosity). The granulometric composition of briquetted material affects not only its compressive strength, but also determines the degree of damage that the brex will undergo under static and dynamic pressure. Depending on the degree of grinding, the fragility of the brex can be replaced with a viscous-plastic destruction, which will, in turn, increase the brex's resistance to impact.

2. Preliminary homogenization of a briquetted batch leads to an increase of its mechanical strength or to a decrease in the amount of binders and plasticizers necessary for its production. It is recommended to use a shearing extruder with a shearing plate to prepare the briquetted batch for homogenization.

3. The flow of a briquetted mass in the extruder is complex and not fully understood. The authors of this paper introduced a simplified model to assess the flow of a compressible mass in an extruder to obtain, for very first time, an exact analytic solution of the Navier-Stokes equations. This approach can be used as a template solution for numerical assessment. Practically all of the known results of mathematical modeling are based on the assumption that a medium is incompressible, which, actually, stands in conflict with the results of this study.

4. The mechanical strength of the brex is affected by their shape. Oval-shaped brex have a different splitting strength depending on the direction at which force is applied to them. The splitting force along the minor axis of the oval is great. The breakdown processes of elongated brex are well described using the SIMULIA Abaqus software, which is based on the finite elements method.

5. In a number of cases, the fines can be carried to the surface of the brex upon their productions. The same phenomenon is well noted in brick production through stiff extrusion. The authors proposed a hydrodynamic model to offer a satisfactory explanation of why a fine fractions may be displaced both in brex and brick production.

6. An analysis of the process of stacking the brex using SIMULIA Abaqus software allows you to model the distribution the brex by orientation in the stack. Knowledge of such distribution allows us to judge the effect of various crushing loads on the brex (axial compression, splitting during stretching).

7. Stiff extrusion agglomeration with the use of cement binder in combination with bentonite, leads to rather high values of impact strength of the brex already at the initial stage of strengthening. The formation of insignificant amounts of fines in brex after three drops from a height of 2 m after 48 h of preparation is due to the viscoplastic nature of their destruction in the first 4 days of strengthening.

8. The possibility of a stiff extrusion line operation on a cement binder is dependent on the necessity to keep the temperature in the room where the briquetting line is located at least 5 °C. Our study showed that after 48-hour strengthening of cement-based wet brex manufactured at 5 °C their drop-strength meet the requirements for blast furnace briquettes.

References

1. Bender, W., Haendle, F.: Brick and tile making, procedures and operating practice in the heavy clay Industries. Bauverlag GMBH, Berlin (1985)
2. Händle, F. (ed.): Extrusion in Ceramics, 468 C. Springer, Berlin (2007)
3. Gregory, R.: Briquetting coal without a binder. Colliery Guardian **201**, 5191 (1960)
4. Kaya, E., Glogg, R., Kumar, S.R.: Particle shape modification comminution. Kona **20**, 185–195 (2002)
5. Ulusoy, U., Hicyilmaz, C., Yekeler, M.: Role of shape properties of calcite and barite particles on apparent hydrophobicity. Chem. Eng. Process. **43**, 1047–1053 (2003)
6. Ulusoy, U., Yekeler, M., Hicyilmaz, C.: Determination of the shape, morphological and wettability properties of quartz and their correlations. Mineral Eng. **16**, 951–964 (2003)

7. Beirne, T., Hutcheon, J.M.: The shape of ground petroleum coke particles Brit. J. Appl. Phys. **13**, 576 (1954)
8. Bizhanov, A.M., Kurunov, I.F., Durov, N.M., et al.: Mechanical strength of BREX: Part I. Metallurg **7**, 32–35 (2012)
9. Bizhanov, A.M., Kurunov, I.F., Durov, N.M., et al.: Mechanical strength of BREX: Part II. Metallurg **10**, 36–40 (2012)
10. Malygin, G.A.: Plasticity and strength of micro—and nanocrystalline materials. Solid State Phys. **49**(6), 961–982 (2007)
11. Kawatra, S.K.: Effects of bentonite fiber formation in iron ore pelletization. Int. J. Miner. Process. **65**, 141–149 (2002)
12. Koizumi, H., Yamaguchi, A., Doi, T., Noma, F.: Fundamental development of iron ore briquetting technology. ISIJ **74**(6), 22–29 (1988)
13. Japan.: Patent S63196689 (A), (1988)
14. Bogdan, E.A., Cole, R.L.: US Patent 5395441, 1995
15. Hideo, K., Nobuhide, O.: Experimental study on swelling characteristics of compacted bentonite. Can. Geotech. J. **40**, 460–475 (2003)
16. Dvorkin, L.I., Dvorkin, O.L.: Building Mineral Binding Materials. Infra Inzheneriya, Moscow (2011)
17. Ozhogin, V.V.: Foundations of Theory and Technology of Briquetting of Pulverized Metallurgical Raw Materials: Monograph. PGTU, Mariupol (2010), 442p
18. Bulatov, A.I., Danishevskii, V.S.: Grouting Mortars. Nedra, Moscow (1987)
19. Jones, G.K.: Chemistry and flow properties of bentonite grouts. In: Proceedings of Symposium on Grouts and Drilling Muds in Engineering Practice, pp. 22–28. Butteworths, London (1963)
20. Shmit'ko, E.I., Krylov, A.V., Shatalova, V.V.: Chemistry of Cement and Binding Substances. Prospekt Nauki, St. Petersburg (2006)
21. Moroz, I.I.: Technology of Structural Ceramics. Ecolit, Moscow (2011), 384p
22. Ruzhinskiy, S., et al.: All about Foam Concrete. 2nd ed., improved and expanded. OOO Stroy Beton (Stroy Beton, LLC), St. Petersburg (2006), 630p
23. Bizhanov, A.M., Eriklintsev, I.V., Kozlov, S.A., Troshkin, O.V.: On Spiral Couette-Poiseuille Flow in Simplified Extruder Problem. J. Comput. Math. Math. Phys. (2017) (in print)
24. Händle, F., Laenger, F., Laenger, J.: Determining the Forming pressures in the extrusion of ceramic bodies with the help of the Benbow-Bridgwater equation using the capillar check. Process Eng. **92**(10–11), 1–7 (2015)
25. Batchelor, G.K.: An introduction to fluid dynamics. Cambridge University Press, Cambridge (1967), 615p
26. Loytsyanskiy, L.G.: Mechanics of Liquids and Gases. Science, Moscow (1987), 840p
27. Abramovich, G.N.: Applied Gas Dynamics. Part 1. Science, Moscow (1991), 600p
28. Bingham, E.C.: Fluidity and Plasticity. McCraw-Hill Book Company, Inc., New York, London (1922), 439p
29. Ishlinskiy, A.Y., Ivlev, D.D.: Mathematical Theory of Plasticity. Publishing House of Physical and Mathematical and Technical Literature, Moscow (2001), 704p
30. Laenger, K.-F., Laenger, F., Geiger, K.: Wall slip of ceramic extrusion bodies, Part 2. Process Eng. **93**(4–5), 1–6 (2016)
31. Belotserkovskiy, O.M., Betelin, V.B., Borisevich, V.D., Denisenko, V.V., Kozlov, S.A., Eriklintsev, I.V., Konyukhov, A.V., Oparin, A.M., Troshkin, O.V.: Toward theory of counterflow in rotating viscous heat-conducting gas. J. Comput. Math. Math. Phys. **51**(2), 222–236 (2011)
32. Troshkin, O.V.: Elements of Mathematical Hydrodynamics and Hydrodynamic Stability. ISBN-978-3-659-93972-3
33. Landau, L.D., Lifshitz, E.M.: Theoretical Physics V.7. Theory of Elasticity. Science, Moscow (1987), 248p

34. Galitskov, S.Y., Nazarov, M.A.: Simulation of Velocity Field of Shear Deformations of Ceramic Mass in Forming Unit of Screw Extruder. Fundamental Studies vol. 8. pp. 29–32 (2013)
35. Joseph, D.: Stability of Fluid Motions. World, Moscow (1981), 638p
36. Mitsoulis, E.: Flows of viscoplastic materials: models and computations. Rheology Reviews 135–178 (2007)
37. Electronic resource http://www.tesis.com.ru Abaqus User Manual, Version 6.12 Documentation
38. Sobolev, A.A., Melnikov, P.A., Tyutyunnik, A.O.: Movement of Particles in Air Stream, vol. 3, Issue No. 17, pp. 82–86. Vector of Science of Togliatti State University
39. Deryagin, B.V., Churaev, N.V., Muller, V.M.: Surface Forces. Science, Moscow (1985), 400p
40. GOST (All-Union State Standard) 2787–75 Metals Ferrous Secondary. General specifications
41. Bizhanov, A.M., Kurunov, I.F., Podgorodetskiy, G.S., Nushtaev, D.V.: Investigation of Mechanism of Brex Destruction under Static and Impact Loads, vol. 8. pp. 26–31. Metallurgist (2014)
42. GOST (All-Union State Standard) 25471-82 Iron Ores, Agglomerates and Pellets. Method for Determining Drop Strength
43. Dorofeev, G.A., Barsukova, E.A.: On the choice of rational method of agglomeration of fine materials of anthropogenic and natural origin. Ferrous Metall. **12**, 73–79 (2015)

Chapter 3
Metallurgical Properties of the Brex and Efficiency of Their Use in a Blast Furnace

3.1 Behavior of the Experimental Brex Under Heating in a Reducing Atmosphere

The debut of SVE technology in blast furnace applications took place in 1993 at Bethlehem Steel (USA). The briquetting line with a capacity up to 20 tons per hour produced brex consisting of basic oxygen furnace (BOF) dust, blast furnace sludge, and coke plant lime sludge. The binder was Portland cement, lignin was used as an extrusion aid. The line operated successfully for 3 years, but was closed once the Bethlehem Steel Corporation stopped operating.

In 2011 a small steel plant of Suraj Products Ltd. (India) started applying SVE technology in their blast furnace unit, and built a line to produce brex using imported converter sludge, blast furnace dust and high-grade iron ore fines. Three years after brex production had commenced they became the main component of the blast furnace charge. At the same time, Novolipetsk Steel Company (JSC NLMK) began to carry out a design study to examine briquetting of blast furnace sludge, blast furnace dust and iron ore concentrate using SVE technology to produce brex for its further use as a blast furnace charge for all the blast furnaces of the company. This method of recycling dispersed iron-bearing waste was a result of the concept of blast furnace, and converter iron–zinc-bearing sludge recycling by briquetting and melting them in Blast Furnace No. 2 with a volume of 1000 m^3, using a special technology allowing zinc to be removed from the blast furnace with blast furnace dust [1–5], which had previously been developed and industrially tested. In connection with the construction of a new large Blast Furnace No. 7 and the decommissioning of Blast Furnace No. 2, use of the converter sludge in the furnace charge for brex production was replaced by a concentrate, since it limited the zinc content in brex and allowed brex to be used in all the blast furnaces at the plant. Blast furnace sludge is normally used in sinter burden, but its amount cannot exceed

© Springer International Publishing AG 2018   69
I. Kurunov and A. Bizhanov, *Stiff Extrusion Briquetting in Metallurgy*,
Topics in Mining, Metallurgy and Materials Engineering,
https://doi.org/10.1007/978-3-319-72712-7_3

6–7 kg/ton of sinter due to high levels of zinc content. However, even this amount of sludge in sinter burden degrades the quality of sinter. During sintering, the fine carbon contained in the sludge does not work as a fuel. Consequently, it is proposed that the sludge be withdrawn from the sintering mixture and used as the charge for brex production.

3.1.1 Reducibility and the Hot Strength Mechanism of Using Brex for Blast Furnaces

We have investigated the metallurgical properties of the brex made from iron–zinc-containing sludge [6, 7], as well as brex made from iron ore concentrate (Stoilensky mining plant, SGOK, Russia) and coke breeze (Tables 3.1 and 3.2) which are considered as promising components of the BF charge, enabling a partial or complete replacement of sinter [8].

Experimental brex were produced by a laboratory computerized extruder, which is able to provide a complete simulation of SVE technology. The brex had a circular section with a diameter of 25 mm and a length ranging between 1.5–2.0 of their diameter. The moisture content of the freshly formed brex was 12.4% (No. 2) and 11.9% (No. 4). The working chamber of the extruder had a vacuum level of 20.2 mm Hg. (No. 2) and 45.72 mm Hg (No. 4). The temperature of the brex when pushed through a die was 30–31 °C.

The compressive strength assessment has been conducted using brex samples with a length of 30 mm. The load was applied along the axis of the samples. The samples were tested before and after the heat treatment in a reducing atmosphere (50% hydrogen + 50% nitrogen) with heating up to 1150 °C at a rate of 500 °C per hour (the temporal temperature gradient is similar to the gradient in a blast furnace shaft), followed by cooling the samples in a nitrogen atmosphere. Five samples of each type were used for the assessment. It was noted that all heat-treated samples kept their shape and dimensions. Despite the significant difference in the component and chemical composition of the two types of brex, i.e., the sludge brex and the brex from concentrate and coke breeze, their strength before heat treatment was practically the same, respectively: 89.4–166.3 kgF/cm^2 (in a sample of 5 brex an average is 119.3 kgF/cm^2) and 103.8–124.7 kgF/cm^2 (in a sample of 5 brex an average is 111.4 kgF/cm^2). After heat treatment in a reducing atmosphere, the strength of the brex decreased, respectively, by 8.2% and by 14.5%. Due to the excess carbon content in brex No. 2 and brex No. 4 and the same reduction mode (the two types of brex were processed simultaneously in the same furnace), the degree of brex metallization (η_{met}) achieved is also practically the same (Table 3.3). At the same time, the atomic ratio of residual carbon content to residual oxygen content (C/O) in metallized brex No. 2 and brex No. 4 is 3.5 and 3.9, respectively. In relation to carbon demand for direct reduction of iron oxides, such atomic ratios indicate excessive amounts of carbon found in the brex in the present study.

Table 3.1 Chemical composition of brex components

Component	Fe_t	FeO	Fe_2O_3	$K_2O + Na_2O$	Al_2O_3	CaO	MgO	MnO	SiO_2	ZnO	S	C
Blast furnace sludge	35.8	11.1	44.6	0.17	0.8	6.4	1.2	0.05	7.8	0.6	0.4	27.3
Blast furnace dust	48.1	9.0	58.7	0.4	0.8	4.45	1.04	0.05	5.9	1–2	0.1	17.0
BOF sludge	56.3	52.0	22.7	–	0.3	12.0	1.3	0.05	2.0	1.5–2.5	0.1	2.3
Iron ore concentrate	66.3	27.9	63.7	0.108	0.18	0.26	0.48	–	7.22	0.003	0.02	–
Coke breeze	1.7	–	–	2.24	3.2	1.0	0.5	–	7.5	–	0.5	85.6

Table 3.2 Components of sludge brex (No. 2) and iron ore concentrate and coke breeze brex (No. 4)

Brex Component	Mass Content, %	
	Brex No. 2	Brex No. 4
Portland cement	9.1	9.0
Coke breeze	–	13.5
Bentonite	–	0.9
Blast furnace sludge	54.5	–
BOF sludge	36.4	–
Iron ore concentrate	–	76.6

However, considering the amount of carbon needed for both direct reduction, and for carburizing iron brex into iron carbide, the excess amount of carbon in brex No. 2 is 2.3%, while brex No. 4 lacks 1.15% to carbonize iron to carbide (when calculating the composition of metallized brex). Thus, the carbon content in the iron ore and the coke-breeze brex appear optimal to meet the carbon demand for direct reduction and its further carburization.

The reducibility study has been performed in accordance with Russian standards —GOST 28657-90 [9] and GOST 21707-76 [10]. The higher rate and higher degree of reduction in iron ore-concentrate-based brex results from the increased oxygen level of iron oxides in this brex (24.6%) compared to the sludge brex (18.1%). Thus, it is possible to conclude that these two types of brex have practically the same reducibility during the process of reduction by gas (with a temperature of up to 900–950 °C, when carbon does not yet take an active part in the reduction process).

These findings allow us to draw an essential conclusion about the desired carbon content in both ore-coke brex and sludge brex (C_{br}). From the point of view of the participation of carbon in the reduction of iron oxides and considering the average degree of direct reduction of iron from partially reduced wustite (r_d) which is equal to 0.4 for the blast furnaces with high rates of fuel injection [11, 12], this content should correspond to a stoichiometric atomic ratio of C/O = 0.3–0.5. With this in mind, and taking into account the presence of metallic iron in brex, the carbon content required for the reduction of the partially reduced wustite (C_{brex1}) should be equal to:

$$C_{brex1} = (0.086...0.1076) \cdot (\%Fe_t, -\%Fe_{met}). \tag{3.1}$$

However, considering that the carbon contained in brex is required not only for reduction of iron from residual wustite, but also for carburization of gas-reduced and carbon-reduced iron, it is proposed the amount of carbon in brex which is sufficient to meet the increased demand for both processes of iron reduction and iron carburization be determined. The carbon content required for carburizing iron to carbide equals $C_{brex2} = 0.0716\%\ Fe_t$. Thus, the total carbon content in brex (ΣC_{brex}) that meets the requirements in carbon for the reduction of iron from wustite, for the carburization of iron, and for the formation of cast iron, is:

Table 3.3 Chemical composition of raw and metallized brex

Brex	Fe_t	Fe_{met}	η_{met}	FeO	CaO	MgO	Al_2O_3	SiO_2	C	Basicity	LOI
No. 2 (Raw)	37.40	–		18.6	15.50	1.26	1.64	7.04	15.7	2.2	19.1
No. 2 (Metallized)	64.5	51.6	0.8	16.8	–			–	9.66		
No. 4 (Raw)	49.20			21.0	6.89	0.58	1.14	8.78	11.6	0.8	12.2
No. 4 (Metallized)	82.6	67.1	0.82	18.9	–			–	7.80		

$$\sum C_{brex} = (0.086...0.1076) \cdot (\%Fe_t, -\%Fe_{met}) + 0.0716\%Fe_t. \qquad (3.2)$$

In comparison with the amount calculated by using the above formula, an excessively high carbon content of coke breeze or coal particles can remain in blast furnace slag. As a result, it can increase slag viscosity and have a negative impact on furnace performance, as mentioned by Bolshakova and Kurunov [13].

Sludge brex and concentrate brex were tested to determine their hot strength properties in accordance with the ISO 4696. The test showed a great advantage of the concentrate brex ($RDI_{+6.3}$ is 96.5%) over the sludge brex ($RDI_{+6.3}$ is 61.8%). These results can be explained by the fact that the concentrate brex contains no hematite, while the sludge brex contains secondary hematite. The presence of secondary hematite in the sludge brex is found in the particles of both blast furnace and converter sludge. The crystal lattice of the hematite rearranges under low temperature and slow reduction, which creates mechanical stresses and disintegration of a hematite-containing material.

For the purposes of a comparative analysis, the hot strength levels of sinters with a basicity (B4) of 1.2, 1.4, and 1.6 produced from a conventional burden used at the NLMK, were simultaneously measured using the same standard. The hot strength level of sludge brex is comparable to that of a sinter with a basicity of 1.2 and 1.4 (64 and 60%), which can be explained by the similar secondary hematite presence in both materials. The hot strength level of a sinter with a basicity of 1.6 (77%) is higher than the hot strength of sludge brex due to the fact that at a basicity of 1.6 or more, sinters enter a new phase by forming calcium ferrite, which strengthens the sinter structure and reduces its disintegration at low-temperature reduction. At the same time, the hot strength level of concentrate and coke-breeze brex significantly outperforms the corresponding characteristics of all tested sinters. This fact is one of the reasons why this brex should be considered as the main component of a blast furnace burden.

The samples of raw and reduced brex were petrographically analyzed on polished sections in reflected light using a Leica DM ILM laboratory microscope with high-resolution optics HC, manufactured by Leica Microsystems. The analysis showed that the structure of the raw brex No. 4 is represented by grains of magnetite, quartz, and coke fines bonded by cement stone—the basis of the strength of raw brex (Fig. 3.1).

The microstructure of the sample of the reduced brex No. 4 (Fig. 3.2) is represented by a number of metallic iron grains sintered with the partially crystallized glass-phase of the iron–olivine composition, and also includes unreacted coke breeze. The strength carrier in the partially reduced brex No. 4 is not only the iron–silicate phase, but also a metallic iron matrix created by sintering metal grains on the surface of the brex.

In addition to particles of sinter, coke, pellets, and slag, the raw sludge brex No. 2 contains ball-shaped metallic beads that are found separately in the cement stone alongside the above-mentioned particles of blast furnace and converter sludge (Fig. 3.3).

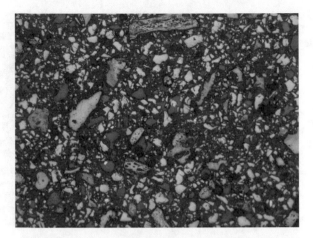

Fig. 3.1 Microstructure of the raw brex No. 4; reflective light; 100 × magnification

Fig. 3.2 Microstructure of the reduced brex No. 4; reflected light; 100 × magnification

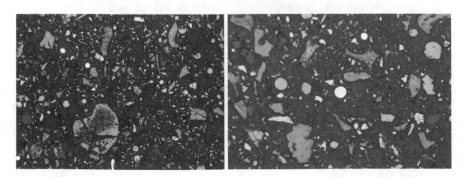

Fig. 3.3 Microstructure of raw sludge brex. Reflected light; 100 × magnification (left); 200 × magnification (right)

Fig. 3.4 Microstructure of
the reduced sludge brex.
100 × magnification

The composition of the reduced brex No. 2 consists of random fine grains of metallic iron, wustite; coke and a silicate phase (see Fig. 3.4). A prominent metal frame, which is observed in the reduced concentrate brex, is absent in the surface layer of the sludge brex. This happens due to lower iron content in the sludge brex, compared to the concentrate brex, and due to a higher content of dispersed coke breeze (Table 3.3).

The silicate phase is the strength carrier of this type of brex, which is similar to brex No. 4. Based on this fact it becomes clear that the formation of a matrix from the iron–silicate phase helps maintain the integrity of the sludge brex when heated in a reducing atmosphere. The integrity in the brex from iron ore concentrate and coke breeze, which has a higher iron concentration, is due to the iron–silicate matrix as well as due to the formation of the metallic iron surface frame formed in the process of gas reduction. The core of the iron ore concentrate brex contains only scattered grains of metallic iron.

For the purpose of a comparative analysis, we studied the metallurgical properties of ore coke brex extruded from hematite iron ore (iron ore—79%, coke breeze—15%, Portland cement—5.55%, bentonite—0.45%). The particles of this high-grade iron ore (Fe_{total}—67.5%; SiO_2—1.5%; Al_2O_3—0.3%; CaO—0.2%; MgO—0.3%; S—0.05%; P_2O_5—0.05%) have a plate-like structure. When used in sinter production, this iron ore shows poor pelletizing ability due to its structure and may downgrade the quality of sinter. However, this very quality of the iron ore does not adversely affect the quality of the brex extruded from it.

As shown in mineralogical studies the major mass of ore minerals is represented by hematite (Fe_2O_3), less often by intergrowths of hematite and magnetite (Fe_3O_4). Silicates are most often found in intergrowths with iron-containing minerals. Figure 3.5 shows the microstructure of the raw brex, the component composition of which is described above.

Fig. 3.5 Microstructure of
the iron ore coke brex;
reflected light; 1—coke
breeze; 2—grains of hematite
iron ore (white) in cement
bond (gray);
50 × magnification

In order to assess the progress of the iron ore coke brex reduction process,
polished sections of brex samples were analyzed after being heated to temperatures
of 900, 1100, and 1200 °C in a reducing atmosphere.

In its center the brex, reduced by heating to a temperature of 900 °C, the
iron-containing phase is mainly represented by wustite and magnetite (Fig. 3.6),
while in the peripheral part of the brex one can see interconnected metallic iron
particles with small silicate inclusions (see Fig. 3.7).

In the peripheral part of the brex sample, heated to 1100 °C, iron oxides get
completely reduced to metal, and one can clearly see a metal frame that has formed
(Fig. 3.8).

Further heating of brex to a temperature of 1200 °C completes the process of
iron reduction to its full extent. The bulk of iron from the central part of brex is
metal and only partially is it wustite. Traces of inclusions of unreacted coke breeze
particles (Fig. 3.9) indicate its excess in the charge used for brex production even

Fig. 3.6 Microstructure of
the central part of the reduced
brex heated to a temperature
of 900 °C; reflected light;
Light gray areas—isolated
areas of reduced wustite and
magnetite; gray areas—
silicate phases;
100 × magnification

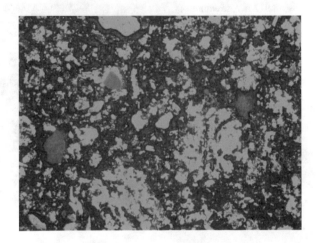

Fig. 3.7 Formation of the
metal framework in place of
large grains of iron ore
hematite (1) on the periphery
of brex (temp = 900 °C);
reflected light;
200 × magnification

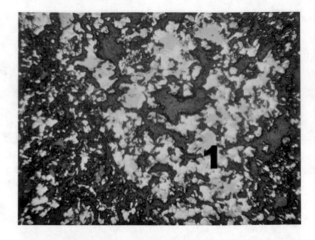

Fig. 3.8 Microstructure of
the periphery of the reduced
brex at 1100 °C; reflected
light; White areas—metal;
gray areas—reduced mineral
phases of the cement binder;
200 × magnification

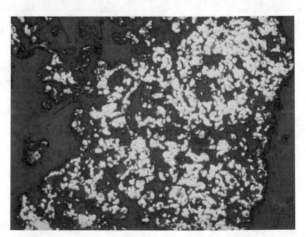

Fig. 3.9 Microstructure of
the center of reduced brex
(at 1200 °C); reflected light;
1—metal; 2—coke breeze;
Gray areas—reduced mineral
phases of the cement binder;
200 × magnification

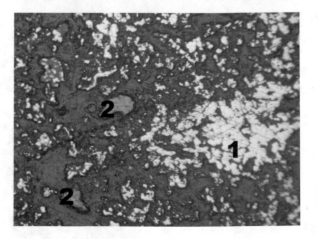

from high-grade ore (67.5% Fe). The carbon content in this brex was 12.9% with an iron content of 53.3%. Taking into account a chemical composition of the brex such as this, according to the stoichiometric equation given above (3.2) the optimal carbon content in it is 8.5–9.5%.

Thus, with the temperature increasing over 900–1000 °C, the coke breeze carbon has a dominant role in the process of iron oxide brex reduction; at the same time, a framework of metallic iron, reduced from oxides by gas, is formed in the peripheral part of the brex. These findings lead to a conclusion that the brex made from anthropogenic materials with low iron content, requires the addition of a natural or anthropogenic component with high iron content. The presence of coke breeze particles in the brex heated in a reducing atmosphere to 1200 °C supports the conclusion that it is necessary to maintain a carbon content in brex in accordance with its stoichiometry in respect of the iron content in the brex.

3.1.2 Metallurgical Properties of Brex Made of Mill Scale and Blast Furnace Sludge

Due to its high iron content, mill scale is the most valuable kind of anthropogenic raw material, which is being recycled at the integrated steel mills with sinter plants by using it in the sintering charge. It is only the oily mill scale that is difficult to recycle. Generally, the oily mill scale is limed and also used in sintering. There are emerging mill scale de-oiling technologies, or technologies in which scale is used as a component of liquid fuel blown into blast furnaces [14] that have no industrial application yet. At the same time, steel mills with no sintering plant face a problem when recycling even oil-free mill scale. The best solution to this problem is to sell scale to third-party consumers. However, one way or another, mill scale can be an effective component of the charge for briquetting in accordance with industrial briquetting technologies, including stiff vacuum extrusion technology. Being sufficiently strong mill scale briquettes can be transported and used as a component of the blast furnace charge; composite briquettes made from scale and coke breeze could be used in electric arc furnaces [15].

Metallurgical properties of the mill scale brex for the blast furnace charge were evaluated with a focus on the mechanical strength and reducibility. Using a laboratory extruder all test brex samples were produced from a mixture of a mill scale (MS) and blast furnace sludge (BFS), with 6% Portland cement (PC) as a binder and 1% bentonite (B) as a plasticizer. The chemical composition of the brex differs primarily in the content of carbon, iron, and oxygen levels (ΣO_2) of iron oxides (see Table 3.4).

We have tested the reducibility of this brex by continuous weighing the samples in a tube furnace in a stream of hydrogen at a temperature of 800 °C. This test showed similar results to those obtained in determining the reducibility of the brex based on a mixture of converter and blast furnace sludge and the brex from iron ore

Table 3.4 Composition of mill scale and blast furnace sludge brex

Brex	Share, %	FeO, %	Fe_2O_3, %	Fe_{tot}, %	$\sum O_2$, %	C, %
B1	MS-80, BFS-13, (PC + B)-7	33.44	49.24	60.47	22.2	3.53
B2	MS-60, BFS-33, (PC + B)-7	27.66	46.30	53.92	20.0	9.03
B3	MS-40, BFS-53, (PC + B)-7	24.10	45.36	50.0	18.96	14.47
B4	MS-20, BFS-73, (PC + B)-7	18.96	43.42	45.30	17.21	19.92

concentrate and coke breeze (Table 3.3 and Fig. 3.1). The reduction rate of the samples of 4 types of brex is directly proportional to the iron and oxygen content of iron oxides in these samples (Table 3.5). Figure 3.10 shows the dynamics of the reduction in the tested brex samples.

The reducibility of the brex from a mixture of mill scale and blast furnace sludge differs significantly due to the different degrees of reducibility of its components. The iron oxides of the blast furnace sludge are mainly found in the sinter particles with a lower degree of reducibility than in mill scale iron oxides.

The near-linear character of the deceleration of brex reduction can be explained by the diffusion-kinetic mode of brex reduction [16], which has a sufficiently high specific volume of pores despite their small diameter. The high carbon concentration in the brex with an increased content of blast furnace sludge at a temperature of 800 °C did not affect the results of the reducibility assessment of the hydrogen-reduced brex, when there was no carbon dioxide (CO_2) in the gas phase and the gasification of carbon present in the brex did not occur. According to the Eq. 3.2 mentioned above, it is clear that brex B3 and B4 had excessive carbon levels, which prevents their use in a blast furnace for reducing oxides of iron, and for carburizing iron within the brex.

These findings enable us to conclude that it appears more efficient to briquette blast furnace sludge in a mixture with a richer iron content and hardly reduced iron-containing components, e.g., in a mixture with a magnetite concentrate. It will contribute to a more effective use of the carbon contained in brex, i.e., the use of carbon in direct reduction and in the carburization of iron, which, as a result, reduces coke consumption in the process.

Table 3.5 Dynamics of the mass loss in hydrogen-reduced brex samples

Brex	The degree of recovery of the samples after the start of the reduction (every 2 min)										
B1	11.4	21.5	30.7	38.0	44.5	49.6	53.8	56.9	59.3	61.3	63.1
B2	10.1	19.5	28.4	36.2	42.7	47.8	51.8	55.1	57.8	59.7	61.3
B3	8.9	16.6	23.1	28.9	33.8	37.6	40.5	42.3	43.8	45.1	46.0
B4	6.7	12.9	18.8	23.5	27.3	30.2	31.4	33.2	34.4	35.9	37.0

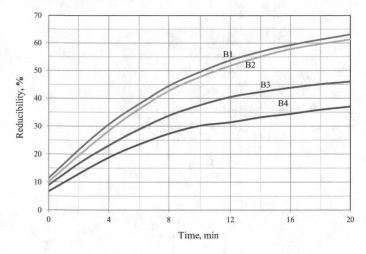

Fig. 3.10 Reducibility of mill scale and sludge brex

3.2 Metallurgical Properties of Industrial Brex for Blast Furnaces

In April 2011 the industrial brex making line was commissioned at the metallurgical plant belonging to Suraj Products Ltd (Rourkela, India). The line produces brex for use in small-scale blast furnaces. The main components of the brex are LD (Linz-Donau) sludge, flue dust, iron ore fines, Portland cement, and bentonite (Table 3.6).

The company's decision was based on the results of the study of the metallurgical properties of brex [17–25] and on the results of the computer simulation of blast furnace smelting performed in a small-scale blast furnace with briquettes in the charge. The main objective of the project was to increase the economic efficiency of blast furnace performance by replacing the purchase of iron ore with cheap brex made of a blast furnace dust and sludge from steelmaking. The layout of Suraj Products Ltd stiff extrusion line is shown in Fig. 3.11.

Green brex are transported by a conveyor belt to a stacking system and unloaded in a pile where they are stored for 2–3 days under natural conditions, and then transported to a blast furnace unit or to outdoor warehouse storage. Due to the high extrusion pressure and vacuumizing in an extruder, the freshly formed brex has a dense structure, high plasticity, and sufficient mechanical strength, which enables the brex to be transported, and stored without disintegration, and the generation of fines. The productivity of the brex making line is 200 t/day with a two-shift operation. In addition to conventional brex used as the main component of a

Table 3.6 Chemical composition of the brex and their constituent components

Brex components shares, %	Fe_2O_3	FeO	Fe_{tot}	LOI	SiO_2	CaO	MgO	Al_2O_3	TiO_2	$K_2O + Na_2O$	C
Iron ore fines 18.9	88.53	–	62.0	4.37	2.71	0.3	–	3.51	0.22	0.2	–
Flue dust 28.3	53.0	–	37.1	–	6.3	4.9	0.2	5.1	–	–	30.5
LD sludge 47.2	17.9	65.4	63.4	–	1.3	8.6	4.9	0.3	–	–	1.5
Portland cement 4.7	4.2	–	–	1.5	20.5	63.2	2.1	4.5	–	0.6	–
Bentonite 0.9	10.0	–	–	9.4	58.4	0.8	0.3	12.6	–	4.4	–
Brex 100	40.4	30.7	52.1	1.4	4.7	8.3	2.5	2.6	0.1	0.13	9.2

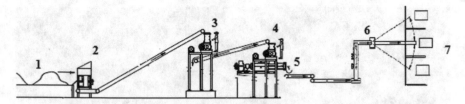

Fig. 3.11 Suraj Products Ltd. (India) stiff extrusion line layout: 1—raw material storage; 2—even feeder with a mixing area for the binder and plasticizer; 3—primary pug-mill; 4—pug-sealer; 5—extruder; 6—piler; 7—strengthening unit

small-scale blast furnace with a working volume of 45 m^3, the brex production unit produces brex from manganese ore fines, which are used to reduce slag viscosity and to improve the drainage capacity of the coke bed.

Before the metallurgical plant started producing brex, its blast furnace unit operated on a charge of 100% rich hematite ore using limestone and dolomite as fluxes.

The study of the physical and mechanical properties of the industrial brex with the composition described above (Table 3.6) has been done in order to comply with all the requirements for charge materials used in blast furnace smelting. These properties of the brex include mechanical strength and porosity, which determine the integrity of the brex during handling and descending in the furnace shaft and their further reduction. The compressive strength of brex has been tested on the Tonipact 3000 equipment (Germany) in accordance with the standard DIN 51067. The apparent porosity has been measured in accordance with the standard DIN 51056 by using the vacuum method of liquid saturation.

To examine pore volume distribution in brex we used Phoenix V|tome|X S 240, an X-ray microtomography system equipped with a 240 kV/320 W micro-focus tube and an 180 kV/15 W high-power nanofocusing tube. Datos|x reconstruction software was used for the primary data processing and for 3D (voxel) model creation from the X-ray images (projections) of the test samples. VG Studio MAX 2.1 and Avizo Fire 7.1 software were used to visualize and analyze the data on elements of the 3D image. The findings from the study showed a uniform pore volume distribution in brex (see Fig. 3.12).

The change in the structure and composition of industrial brex heated to 1500 °C in a CO atmosphere has been studied in comparison with the change in the structure and composition of the rich hematite iron ore. The samples of both materials were heated in a reducing atmosphere simultaneously. When a temperature of 1000, 1100, 1200, 1300, 1400, and 1500 °C was reached the samples were drawn to determine the reduction and the degree of metallization in the ore and brex. The duration of each test was 2 h. The chemical composition of the samples of iron ore and brex is given in Table 3.7.

Fig. 3.12 Distribution of pores larger than 5 μm in brex volume

Table 3.7 Chemical composition of iron ore and brex, %

Elements	Iron ore	Brex
Fe_2O_3	88.53	40.4
FeO	2.85	30.7
SiO_2	2.71	4.7
CaO	0.3	8.3
MgO	–	2.5
Al_2O_3	3.51	2.6
TiO_2	0.22	0.1
$K_2O + Na_2O$	0.20	0.13
C	–	9.2
LOI	1.6	1.4

An X-ray diffractometer ARL 9900 XRF was used to identify the phase composition of brex charge components, which includes:

– Iron ore—hematite, goethite, gibbsite, pyroxene, kaolin;
– blast furnace dust—magnetite, hematite, wustite, graphite, quartz, dicalcium ferrite;
– LD sludge—magnetite, wustite, wollastonite, calcite.

The reduction rate of the samples was determined by measuring the weight loss in the sample during the reduction process and by the initial oxygen content of iron oxides in the samples.

$$\text{Reduction degree } (\%) = \left(\frac{M_i - M_f}{M_o}\right) \times 100$$

Table 3.8 Degree of metallization and reducibility in iron ore and brex

T, °C	Fe_{tot}, %		Degree of metallization, %		Reducibility, %	
	Iron ore	Brex	Iron ore	Brex	Iron ore	Brex
1000	67.5	60.5	3.2	23.5	35.0	48.0
1100	69.5	61.7	18.3	33.0	52.0	62.0
1200	70.3	63.7	25.1	47.6	61.0	72.0
1300	73.3	65.6	41.3	68.0	65.8	80.0
1400	80.8	67.5	66.5	77.2	82.0	88.1
1500	90.4	80.9	93.2	97.0	93.0	99.0

M_i —sample weight before heating, g
M_f —sample weight after heating, g
M_o —the total amount of oxygen in the iron oxides of the initial sample, g

The degree of metallization in iron ore samples and brex samples was measured by the ratio of the metallic iron content to the total iron content. At all temperatures, the reducibility and metallization of the brex samples exceeded the reducibility and the degree of metallization of the iron ore (Table 3.8). The higher reducibility of the brex during the reduction at a temperature ranging from 500 to 1000 °C can be explained by the higher porosity of the brex (33.7% at 500 °C and 37.4% at 1000 °C) in comparison with the porosity of the ore (19.4 and 29.2%, respectively). Moreover, at temperatures above 1000 °C, the higher rate of oxygen removal from the iron oxides in brex is due to the carbon presence in the brex. The carbon is actively involved in the process of reduction of iron oxides at the above-mentioned temperatures [26].

The reduced samples of industrial brex were analyzed petrographically by using a Leica DM ILM laboratory inverted microscope with high-resolution optics HC produced by Leica Microsystems (Germany). In addition, the samples were studied using scanning electron microscopy methods using an ECLIPSE LV100-POL polarized microscope equipped with a DS-5M-L1 digital photomicrography system.

The structure of the brex sample that was reduced at 1400 °C (Figs. 3.13 and 3.14) is represented by the metallic iron grid with the micro zones in the wustite. The metal is released from the glass phase, which has been recrystallized to silicates. Continuous fields of metal and differently colored glass phases are observed.

Thus, the integrity of industrial brex heated in a reducing atmosphere after the loss of cement stone strength is achieved by the formation of a surface metal frame during the reduction of iron oxides by a gaseous reducing agent. In addition, the formation of an iron–silicate matrix, with iron reduced by coke breeze carbon at temperatures above 1100 °C, contributes to the integrity of the brex.

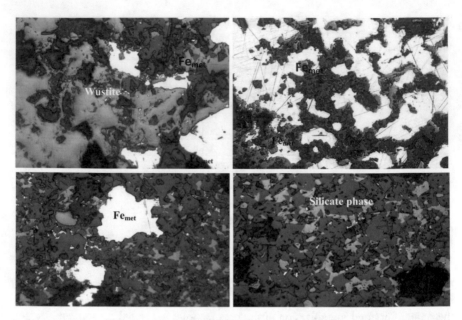

Fig. 3.13 Brex microstructure when heated in a reducing atmosphere at 1400 °C (200 × magnification)

Fig. 3.14 Focused-beam microscopy of brex after reducing heat (1400 °C) 1—α-iron, 2—wustite, 3, 4—glass phase, 5—wustite

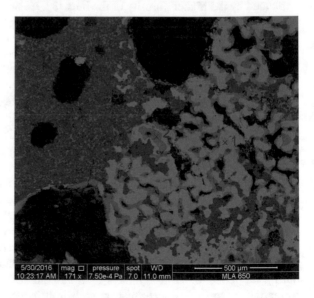

3.3 Blast Furnace Operation with 100% Brex Charge

The small-scale blast furnace (volume 45 m³) at Suraj Products Ltd is equipped with a skip hoist with a volume of 0.5 m³, a double-cone charging device, and hydraulic equipment for notch servicing, recuperative stoves and a two-stage dry gas-cleaning system (dust collector + the seven modules of the bag house filters) and sludge granulation equipment. The furnace has external watering and eight air tuyeres. The hot metal (foundry and pig iron) is immediately casted by the casting machine. The daily production pf the blast furnace is 100–130 tons of pig iron.

The extrusion line of the blast furnace unit began operating in May 2011, using 10% of brex in the blast furnace charge. Gradually, the brex content of the charge increased to 80%. From the end of 2013, the blast furnace has been operating using 100% brex charge. Table 3.9 presents the main furnace operating parameters with brex in the charge at Suraj Products Ltd.

Table 3.9 Furnace operating parameters at Suraj Products Ltd.

Furnace operating parameters	100% ore	80% brex	100% brex
Consumption, kg/t			
Iron ore	1500	372	–
Brex	–	1425	1960
Limestone	150	–	–
Dolomite	144	–	29
Scrap	132	–	–
Quartzite	–	–	13
Mn ore brex	–	19	75
Coke[*]	680	530	490
Fe_{tot} in charge, %	57.6	50.4	45.5
Productivity, t/m³ per day	1.9	1.62	2.0
Blast temperature, °C	925	900	1000
Blast pressure, kg/cm²	0.5	0.34–0.38	0.38–0.42
[Si], %	1.0–1.8	1.0–1.5	0.8–1.1
[Mn], %	0.2	0.4–0.5	0.7–0.8
[C], %	3.8–4.0	3.75–3.90	3.80–3.95
[S], %	0.050–0.060	0.038–0.050	0.038–0.042
Cast iron temperature, °C	1380–1440	1400–1450	1410–1450
(CaO), %	34.86	33.12	38.0–39.0
(SiO_2), %	31.98	30.23	30.0–32.0
(Al_2O_3), %	23.87	17.98	16.0–18.8
(MgO), %	9.46	9.48	8.0–9.5
(FeO), %	1.01	1.26	0.6–1.15
(MnO), %	0.35	0.75	1.3–1.0

[*]The size is 15–25 mm

The data shows that when working on the charge made up of 80% brex and 20% iron ore coke consumption decreased by 150 kg per ton of hot metal (22%) compared with operation on 100% iron ore. The decrease in coke consumption was due to the carbon contained in brex and its involvement in the iron reduction process, as well as to the removal of limestone and dolomite from the charge. The productivity of the blast furnace that used 80% of brex in the charge decreased mainly due to a 7.2% decrease in iron content in this charge compared with the charge based on iron ore and raw fluxes. The transition to a mono-charge blast furnace operation resulted in a further decrease in coke consumption due to the additional carbon input. The increase in the blast temperature by 100 °C also contributed to the decrease in coke consumption. In addition, the use of "washing" brex from manganese ore reduced the sludge viscosity and improved the processing of smelting products. As a result, it contributed to the blast furnace performance gain along with the improvement in the structure of the stock column and the increase in the overall pressure drop. The decrease in coke consumption was also a result of specific heat loss reduction.

The development of new of blast furnace smelting technology with the use of brex meant switching to a decreased stock level of blast furnace charge due to the difficulties in the operation of dry gas-cleaning system. The gradual increase in the share of brex in the charge led to a significant increase in the moisture content of the furnace top gas due to the added hydrated moisture resulted from the brex in the blast furnace and due to a drop in the temperature of the top gas below the dew point. The moisture condensation on the bag filters caused them to become clogged with finely dispersed top dust. As a result, the filters could not be regenerated by inverse pressure pulses. Lowering the burden level enabled an increase in the temperature of the top gas and stopped the bag filters from sealing. This change in burden level had no significant effect on the blast furnace performance due to the fact that the lower iron oxide (FeO) in brex gets reduced directly by the carbon from brex coke; meanwhile, the high reduction potential of the hearth gas and its temperature allows for a fast reduction of iron oxides to wustite.

Despite high specific heat losses due to the small volume of the blast furnace and low blast temperatures, the blast furnace coke consumption does not exceed 500 kg/ton when using the charge with brex. Similar results are obtained by modern high-efficiency blast furnaces that use a charge containing 58–59% of iron, and have a blast temperature of 1200 °C with an excessive top gas pressure of 180–250 kPa. This is achieved by the self-reducing nature of brex which contains carbon from furnace flue dust. Finely dispersed coke particles in brex are in close contact with dispersed particles from steelmaking sludge and iron ores, which provides a direct reduction of iron oxides without the participation of coke carbon. In addition, due to the high basicity of LD sludge and cement, the resulting high basicity of brex (B2 = 1.76, B4 = 1.48) requires no raw limestone and dolomite to obtain a specified basicity of blast furnace slag. To adjust the slag basicity and to optimize the content of magnesia in the slag, it is necessary to modify blast furnace charge by adding small quantities of quartzite and dolomite. As a result, using the 100% brex charge reduced the coke consumption for iron smelting by almost 200 kg/ton compared to the blast furnace operation with 100% rich iron ore, with limestone and dolomite.

The results obtained during over 6 years of the industrial brex making line operation support the fact that the brex made from natural and anthropogenic dispersed raw materials has a controlled and optimum size, a controlled chemical composition, high metallurgical properties, which makes the brex a fundamentally new component of blast furnace charge. Unlike sinter and pellets production, brex production is environmentally friendly and completely waste free, with no gaseous or solid emissions. There are practically no fines left after loading and unloading brex into the blast furnace (see Fig. 3.15). When using a single-component charge consisting of 100% brex the small-scale blast furnace demonstrates uniquely low coke consumption (490–500 kg/t) at a blast temperature of 1000 °C and with an iron content of 45.5% in the fluxed charge.

Fig. 3.15 Brex in blast furnace skip at Suraj Product Ltd.

3.4   Assessment of Potential Usage of Brex Made of Iron Ore Concentrate with Carbon Addition in Blast Furnaces

Preliminary qualitative assessments of the possibility and efficiency of brex used on a large scale as a new component of blast furnace charge were based on both the research conducted into the metallurgical properties of brex from various components, and on the unique experience of using a single-component brex charge in a small-scale blast furnace. This assessment was completed by using the numerical simulation of a blast furnace process, based on the DOMNA computer program developed by I.F. Kurunov, S.B. Yaschenko and L.A. Fursova in National University of Science and Technology MISiS (Moscow, Russia) [27]. The process of blast furnace smelting was simulated in a furnace with a volume of 4297 m^3, operating at Novolipetsk Steel, NLMK. The process of blast furnace smelting using a charge from three components, i.e., sinter, pellets and brex were modeled with respect to the brex basicity, the level of which depends on the proportions of cement and bentonite used as a binder and plasticizer. The pellet content is determined by the capacity of the Stoilensky GOK pelletizing plant built by NLMK (6 million tons of pellets per year). The basicity of the brex made of iron ore concentrate and caking coal depends on the content of cement (6%) and bentonite (1%) in the brex charge mixture, the content that was determined by experiments on brex from this charge mixture, and by the coal content in the charge mixture. The basicity of the brex (B2) with the above-mentioned cement and bentonite contents is 0.50–0.55. The basicity of sinter is determined in accordance with an accepted approach of replacing sinter with brex and brex basicity. When replacing 50% of sinter by brex in the blast furnace mixture, the basicity of the sinter must be within a range of 2.8–3.2. It should be noted that with this basicity, calcium ferrites predominate in the structure of the sinter, and increase sinter strength in comparison with the basicity of 1.5–1.7 in the sinter produced at NLMK [12].

Rated sinter compositions with the basicity of 1.70 and 3.02, brex from the SGOK concentrate and caking coal with two different carbon contents (see Table 3.10) were used to simulate different variants of blast furnace operation. The carbon content in brex B1 is determined by the Eq. (3.1), i.e., it corresponds to the need for a carbon reducing agent with a degree of direct reduction that is equal to 40%, and corresponds to the degree of direct reduction of 30–50%, which is observed in the intensively operated blast furnaces of Novolipetsk Steel with a high consumption of injected natural gas [12, 28, 29].

The carbon content in brex B2 is determined by the correlation (3.2), and corresponds to the total need for carbon for the process of iron reduction from wustite, and for the carburization of iron, i.e., for cast iron production from the iron contained in brex.

The simulation was carried out for the same composition of hot metal and its temperature ([Si] = 0.4%; [C] = 4.8%; T$_{iron}$ = 1500 °C) and for the same reduction efficiency (the degree of approximation to the equilibrium composition of the gas in

Table 3.10 Chemical composition of raw materials, sinters and brex, %

Materials	Fe_{tot}	FeO	Fe_2O_3	SiO_2	Al_2O_3	CaO	MgO	C	SO_3
Portland cement	3.3	–	4.71	20.64	4.98	63.58	1.15	–	2.55
Bentonite	3.45	0.5	4.37	59.25	14.27	2.07	3.62	–	0.14
Caking coal	0.84	–	1.2	2.7	1.5	0.4	0.1	68.9	0.36
SGOK pellets	64.38	1.51	90.31	7.04	0.32	0.22	0.45	–	0.12
SGOK Iron ore concentrate	66.32	29.2	62.3	6.62	0.18	0.26	0.1	–	0.05
Sinter (170)	55.4	11.81	66.0	6.70	0.72	11.37	2.51	–	0.05
Sinter (3.02)	50.34	10.0	60.8	6.30	0.7	19.0	3.0	–	0.06
Brex B1	56.6	24.83	53.28	7.67	0.71	4.09	0.19	5.5	0.22
Brex B2	54.34	23.84	51.15	7.36	0.68	3.93	0.18	9.5	0.21

Table 3.11 Results of the blast furnace smelting simulation when using a traditional charge and a new charge with one-third iron–carbon-containing brex

BF operation parameters	Basic variant	Variant 1	Variant 2	Variant 3
Sinter consumption B2 = 1.7, kg/t	1109	–	–	–
Sinter consumption B2 = 3.0, kg/t	–	557	575	575
SGOK pellets consumption, kg/t	546	557	541	565
Brex consumption B1, kg/t	–	557	575	–
Brex consumption B2, kg/t	–	–	–	575
SGOK iron ore consumption, kg/t	–	17	–	–
Fe content in charge, %	58.2	57.45	57.15	56.2
Coke rate, kg/t	391	354	284	257
Natural gas consumption, nm^3/t	125	125	35	35
Pulverized coal consumption, kg/t	–	–	160	160
Blast rate, m^3/min	7483	7568	7340	7340
Blast temperature, °C	1240	1240	1240	1240
O_2 content in blast, %	30.5	30.5	30.5	30.5
Blast humidity, g/m^3	10	10	20	20
Top gas yield, m^3/t	1545	1540	1470	1466
Top gas pressure, kPa	240	240	240	240
CO, %	24.4	24.9	26.2	26.6
CO_2, %	23.2	22.6	23.9	24.4
H_2, %	9.7	9.9	8.2	8.3
Slag ratio, kg/t	318	314	323	320
Slag basicity, B2	1.01	1.01	1.02	0.99
Capacity, t/day	12,465	12,624	12,708	12,825
Capacity, $t/m^2 \cdot day$	92.48	93.66	94.3	95.2
Reduction efficiency, %	94.2	94.2	94.2	94.2

the wustite reduction zone). The results of the experiment showed that at a brex consumption rate of 557 kg per ton of hot metal, each 1% of carbon in brex [carbon content is calculated by the dependencies (3.1) and (3.2)], reduces coke consumption by 6.7 kg/ton. The blast furnace that uses B1 brex and 125 m^3 of natural gas per ton of hot metal consume less coke, the amount of which decreases from 391–354 kg/t in comparison with the basic variant. When pulverized coal is injected in an amount of 160 kg/t and 35 m^3/t of natural gas is used, brex B1 reduces coke consumption to 284 kg/t (see Table 3.11).

3.5 Conclusions

1. The cement–bentonite binder in iron ore and sludge brex provides sufficient mechanical strength, which simplifies the transportation of green brex using a conveyor belt to a stacking system without disintegration or fines generation. After 48 h of strengthening brex is strong enough to be loaded into wagons for further transportation to a blast furnace unit and to be loaded into the furnaces bins without forming fines as well. The viscous-plastic character of brex destruction is maintained during compression tests performed in the course of 4 days of strengthening.
2. The level of hot and cold strength of the types of brex that were investigated meets the requirements for the components of a blast furnace charge. The hot strength of brex from anthropogenic raw materials is comparable to the hot strength of fluxed sinter. The hot strength of brex from magnetite concentrate and coke breeze is much higher than the hot strength of fluxed sinter.
3. The formation of an iron–silicate matrix helps maintain the integrity of sludge brex when heated in a reducing atmosphere to 1150 °C. In the brex from iron ore concentrate and coke breeze in addition to the iron–silicate matrix formation a metallic iron surface frame is also formed in the process of reduction.
4. Effective use of carbon which is contained in brex requires carbon levels to be below or equal to the amount determined by the stoichiometric dependence of carbon content in brex on the iron content within brex: $C_{br} = (0.086...0.1076)$ $(Fe_t - Fe_{met}) + 0.0716Fe_t$.
5. The reducibility of brex from a mixture of mill scale and blast furnace sludge in the hydrogen atmosphere depends on the share of blast furnace sludge, decreasing as its increases.
6. Industrial brex from blast furnace dust, converter sludge, and iron ore fines with a cement–bentonite binder have a sufficiently high and evenly distributed porosity, which provides for their high reducibility.
7. Many years of successful operation of a small blast furnace with the charge of 100% of the brex made from a mixture of natural and anthropogenic raw materials with a blast temperature of 1000 °C and with the coke consumption

smaller than 500 kg/t enables a conclusion to be made that brex are a new effective component in the BF charge and can be considered as a partial, or complete alternative to sinter.

8. The results of the numerical simulation of blast furnace smelting using brex from iron ore concentrate and caking coal with a brex carbon content of 9.5% showed a high efficiency of partial (by 50%) replacement of a sinter production which can lead to reduction of coke consumption by 15% and to a 50% reduction of gaseous and dust emissions during sinter production.

References

1. Lisin, V.S., Skorokhodov, V.N., Kurunov, I.F., Chizhikova, V.M.: Current state and prospects of recycling of zinc-containing wastes of metallurgical production. Ferrous Metallurgy. Bulletin of Institute of Technical and Economic Research and Information in Ferrous Metallurgy. Appendix 6, p. 32 (2001)
2. Kurunov, I.F., Kukartsev, V.M., Yarikov, I.S., Emelyanov, V.L., Titov, V.N.: Industrial recycling of residues containing iron and zink by sintering and smelting them in blast furnace (the experience of OAO NLMK (Novolipetsk Metallurgical Combine, OJSC)/. Steel. 10, 15 (2003)
3. Kurunov, I.F, Grekov, V.V, Yarikov, I.S, Grigoryev, B.N., Kuznetsov, A.S., Emelyanov, V.L., Titov, V.N.: Production and smelting in blast furnace of agglomerate from residues containing iron and zink. Ferrous Metallurgy. 9, 33 (2003)
4. Lisin, V.S., Skorokhodov, V.N., Kurunov, I.F., Chizhikova, V.M., Samsikov, E.A.: Resource and ecological solutions for disposal of wastes from metallurgical production. Ferrous Metallurgy. 10, 64 (2003)
5. Kurunov, I.F., Titov, V.N., Bolshakova, O.G.: Analysis of effectiveness of alternative ways of recycling metallurgical wastes containing iron and zink. Metallurgist 11, 39–42 (2006)
6. Patent of the Russian Federation No. 2506327 Application of 18.05.2012 Published on 10.02.14 Bulletin No. 4
7. Patent of the Russian Federation No. 2506326 Application of 18.05.2012. Published on 10.02.14 Bulletin No. 4
8. Patent of the Russian Federation No. 2506325 Application of 18.05.2012 Published on 10.02.14 Bulletin No. 4
9. GOST (All-Union State Standard) 28657–90: Iron Ores. Method for Determining Reduction
10. GOST (All-Union State Standard) 21707-76: Iron Ores, Agglomerates and Pellets. Method for Determining Gas Permeability and Layer Shrinkage During Reduction
11. Blast Furnace Process Reference Guide: Metallurgy, vol. 1, Moscow (1989)
12. Filatov S.V., Kurunov I.F., Tikhonov D.N., Basov V.I.: Influence of smelting intensity on blast furnace productivity and coke consumption. Metallurgist. 7, 20–24 (2016)
13. Kurunov I.F., Shcheglov E.M., Kononov A.I., Bolshakova O.G. et al.: Investigation of metallurgical properties of briquettes from technogenic and natural raw materials and assessment of effectiveness of their application in blast furnace smelting. Part 1. Bulletin of Scientific and Technical and Economic Information "Ferrous Metallurgy" 12, 39–48 (2007)
14. Kurunov, I.F., Petelin, A.L., Tikhonov, D.N., Erokhin, S.F.: Injection of combined liquid fuel from oil wastes and oily scale into blast furnace. Metallurgist 7 (2004)
15. O'Kane, P., Fontana, A., Skidmore, K., Jin, Z.: Sustainable steelmaking through the use of polymer technology. Iron Steel Technol. 1, 88–96 (2017)

16. Vegman, E.F., Zherebin, B.N., Pokhvisnev, A.N., et al.: Cast iron metallurgy, p. 774. Textbook for Higher Institutions. Publishing and Book Trade Centre "Academical Book", Moscow (2004)
17. Bizhanov, A.M., Kurunov, I.F., Durov, N.M., et al.: Mechanical strength of BREX: Part I. Metallurgy. **7**, 32–35 (2012)
18. Bizhanov A.M., Kurunov I.F., Durov N.M., et al.: Mechanical strength of BREX: Part II. Metallurgy **10**, 36–40 (2012)
19. Dalmia, J.K., Kurunov, I.F., Steele, R.B., Bizhanov, A.M.: Production and smelting briquettes of new generation in blast furnace. Metallurgist. **3**, 39–41 (2012)
20. Kurunov, I.F., Bizhanov, A.M., Tikhonov, D.N., Mansurova, N.R.: Metallurgical properties of brex. Metallurgist. **6**, 44–48 (2012)
21. Kurunov, I.F., Bizhanov, A.M.: Brex is new stage in sintering raw materials for blast furnaces. Metallurgist. **3**, 49–53 (2014)
22. Bizhanov, A.M., Kurunov, I.F., Dashevskiy, V.Y.: On mechanical strength of extrusion briquettes. I. Metals. **2**, 19–25 (2015)
23. Bizhanov, A.M., Kurunov, I.F., Dashevskiy, V.Y.: On mechanical strength of extrusion briquettes. II. Metals. pp. 3–10 (2015)
24. Bizhanov A.M., Kurunov I.F., Ivonin D.V. Mechanism of fine fraction migration at extrusion briquettes (Brex) production. Metallurgist. **7**, 24–28 (2013)
25. Bizhanov, A., Kurunov, I., Dalmia, Y., Mishra, B., Mishra, S.: Blast furnace operation with 100% extruded briquettes charge. ISIJ Int. **55**(10), 175–182 (2015)
26. Matsui et al.: ISIJ Int. **43**(12), 1904M
27. Kurunov, I.F., Yaschenko, S.B.: A calculation method for the technical and economic parameters of a blast furnace. Scientific works of the Moscow Institute of Steel and Alloys. # 152 (1983)
28. Kurunov, I.F., Filatov, S.V., Bizhanov, A.M.: Assessment of effectiveness of use of ore-coal brex in smelting in blast furnace by mathematical modeling. Metallurgist. **10**, 23–25 (2016)
29. Geerdes, M., Chaigneau, R., Kurunov, I., Lingiardi, O., Rikkets, J.: Modern Blast Furnace Ironmaking. IOS Press BV, Amsterdam (2015)

Chapter 4
Metallurgical Properties of Brex and the Efficiency of Their Use in Ferroalloys Production

Significant volumes of finely-dispersed, effectively-recyclable materials are created in the process of extraction and beneficiation of raw materials for ferroalloy production during their transportation to ferroalloy plants and ferroalloy smelting.

In particular, large reserves of manganese ore with below commercial grade manganese content have been stored at mining and processing plants. When additionally ground, this material becomes easy to process by magnetic and gravitational beneficiating methods, with the possibility of achieving a manganese content of 35–40% Mn in the concentrate.

One of the most common types of anthropogenic raw materials resulting from ferroalloy production are dust and sludge, obtained in gas cleaning of ferroalloy furnaces. Dust emissions from ferroalloy furnaces occur at a rate 8–30 kg/t of ferroalloys. The manganese oxides content of dust formed in the production of silicomanganese and ferromanganese are 21.0–34.3% and 30–35% respectively. The chromium oxide content of dust formed in the production of ferrochrome is 22–30% [1]. The use of finely-dispersed raw materials without their agglomeration in submerged electric arc furnaces (SEAF) disrupts the normal course of the process leading to increased dust generation, reduced productivity and metal extraction rate, increased risks of accidents and hot-idle hours, and as a result, an overall decrease in the technical and economic performance of the furnace.

Sintering is a traditional method of agglomerating such finely-dispersed materials. Ferroalloy factories did not have sintering plants until 1973 [2]. The first sintering plant was opened at Nikopol Ferroalloy Plant in 1973 (Ukraine). In 1979, a sintering plant was commissioned at the Zestafoni Ferroalloy Plant. The average productivity of sinter machines is 100 t/h (specific—1.2 t/m^2 per hour).

Another industrial method of agglomeration is pellet calcination at a temperature of 1150–1220 °C. Pellet strengthening is ensured by the formed liquid phase. At the "Kashima" plant (Japan), for high-carbon ferromanganese smelting in SEAF with a capacity of 40 MVA, manganese concentrates are agglomerated in a 110 t/h capacity pelletizer and then calcined in a 75 m long rotary kiln with a diameter of 3.5 m [3].

© Springer International Publishing AG 2018
I. Kurunov and A. Bizhanov, *Stiff Extrusion Briquetting in Metallurgy*,
Topics in Mining, Metallurgy and Materials Engineering,
https://doi.org/10.1007/978-3-319-72712-7_4

The industrial application of roll briquettes for the agglomeration of anthropogenic and natural raw ferroalloy industry materials was described in Chap. 1.

The first successfully implemented project with the use of SVE technology in the ferroalloy industry was the production of brex and their use in the charge for ferronickel smelting at BHP Billiton in Cerro Matoso (Montelibano, Colombia [4] in the framework of the RKEF (rotary kiln—electric furnace) process [5]. The nickel content in this ore is usually small and ranges from 0.8 to 3.0%. During the RKEF process, laterite ores are sieved, ground and added into a charge with a certain iron, nickel, SiO_2 and MgO content. After that, the charge material is roasted and partially metallized with coal or coke fines. This semi-product and residual coke are then delivered to the electric furnace where ferronickel is smelted. In Cerro Matoso, the RKEF process entails the use of briquetted fines of nickel laterite ore.

Initially, the plant, which has been in operation since 1992, used pelletization and roll briquetting to agglomerate fines of nickel laterite ore and dust procured through the gas cleaning of electric furnaces. In 1996, after a number of comparative tests of various agglomeration and processing methods for these materials (pelletizing, agglomeration, plasma furnaces and SVE), the final choice was made in favor of SVE technology. This technology allowed for an agglomerated product with a cold and hot strength sufficient to maintain brex integrity when roasted in tube kilns (1150 °C) and thermal stability when smelted in electric furnaces. Moreover, SVE allowed for high productivity while maintaining minimal production cost. BHP Billiton Ltd commissioned three Steele 90 Series extruders to produce 700,000 tons of brex per year from laterite nickel ore fines and gas cleaning dusts for the further metallization of brex in rotary tube kilns and then for its use as a component of the charge for ferronickel smelting in electric furnaces. When using the SVE technology for sintering laterite nickel ore fines and dust, it becomes possible to obtain strong brex without the use of a binder. The production of nickel from ferronickel by BHP Billiton in Cerro Matoso is 50,000 tons per year, including 10,000 tons attributed to brex. The nickel content in commercial materials is 35% [6]. The successful results of this industrial use of the SVE technology attracted the attention of VALE, one of the largest mining companies in the world. This Onca Puma-based company (Brazil) has commissioned three extrusion lines for agglomeration of laterite nickel ore fines, with an annual brex production of 700,000 tons.

SVE technology had not been used previously to agglomerate manganese and chromium ores. In 2010, we were the first to attempt to use SVE technology for such purposes. We planned to conduct a full-scale pilot testing of this technology using brex in the charge of industrial SEAF. The rationale for the pilot study was to select the optimal method for using manganese ore fine and for recycling dry gas cleaning manganese-containing dust generated during the production of silicomanganese. The possibility of applying the SVE method to the production of ore brex and ore and dust brex has been studied.

4.1   Brex from Manganese Ore Concentrate

Brex from primary oxidized manganese ore concentrate produced at the Zhairem mining plant (Kazakhstan Republic) was tested for the purpose of this study. The brex was manufactured on the industrial SVE line. Portland cement (5% of weight) was used as a binder; bentonite (1% of weight) was used as a plasticizer. The appearance of the brex at the exit of the die is shown in Fig. 4.1.

Table 4.1 shows the chemical composition of the primary oxidized manganese concentrate of Zhairem mining plant.

The share of particles larger than 0.071 mm is 22%; the manganese content of these particles is 35.4%. The share of particles smaller than 0.071 mm is 78%; the manganese content of these particles is 38.77%. Iron particles distributed in the samples of the concentrate in both size ranges evenly—3.81–3.86%. The surface area of cement particles exceeds 4000 cm^2/g. The chemical composition of cement is: CaO—62–64%; MgO—2.5% (max.); SiO_2—22–24%; Al_2O_3—4.5%; Fe_2O_3—4.5%; SO_3—1.8%; alkali—0.6%. The chemical composition of bentonite (%) is: SiO_2—58.4; Al_2O_3—12.6; CaO—0.8; Fe_2O_3—10.6; alkali—4.4.

Fig. 4.1 Brex from primary oxidized manganese ore concentrate

Table 4.1 Chemical composition of primary oxidized manganese concentrate of Zhairem mining plant

Elements	Mn	Fe	SiO_2	Al_2O_3	CaO	MgO	Ba	Na_2O	P	S	Zn
Concentrate, %	38	3.81–3.86	10–11	1.36	15–19	0.76	0.39	0.18	0.05–0.09	0.09–0.12	0.084

Physical and mechanical properties of the brex were measured after three-day strengthening period in an open warehouse. The compressive strength amounted to 30 kgF/cm² and was measured on the Tonipact 3000 equipment (Germany) in accordance with the DIN 51067 standard. The open porosity was determined in accordance with DIN 51056 and amounted to 14.39%. The density of the brex was an average of 2.59 g/c³.

A specially designed unit was used to study the behavior of industrial brex samples from primary oxidized manganese concentrate from the Zhairem mining plant during heating heated in a reducing atmosphere. The test chamber was a hollow corundum tube with an outer diameter of 85 mm and an inner diameter of 75 mm. The brex samples were placed in a graphite crucible placed inside the chamber and covered with coke breeze particles with 1–5 mm in width. The thickness of the coke breeze layer covering the entire crucible was 25 mm. After that, perforated inserts were used to block the crucible from both ends, which provided heat insulation. Water-cooled metal clips were installed on both sides of the corundum tube. The vacuum pump was connected to one of the clips together with an air supply system and a flow meter. Before heating, the internal volume of the corundum tube was evacuated to 0.5 mbar. Then, the heating process started (10 °C per minute) and after the temperature reached 800 °C, the air supply was switched on at a flow rate of 10 L per minute until the end of the test. The brex samples were subjected to reducing roasting (in the CO atmosphere) in 1000–1500 °C temperature range at intervals of 100 °C. The test was performed at a fixed temperature for the duration of 2 h. The installation diagram and the sample positioning method are shown in Figs. 4.2 and 4.3.

The physical and mechanical properties of brex after heating at different temperatures are given in Table 4.2.

The Rigaku MiniFlex600 X-ray diffractometer was used to perform a phase analysis of original brex samples and the brex samples that were heated in a reducing atmosphere for 2 h. The main phases are presented in Table 4.3.

Fig. 4.2 Installation diagram to study brex behavior when heated in reducing atmosphere. 1—oven, 2—controller, 3—hollow corundum tube with samples, 4—vacuum pump, 5—compressor, 6—isolation valve, 7—drying device, 8—flow meter, 9—gate valve, 10—burner

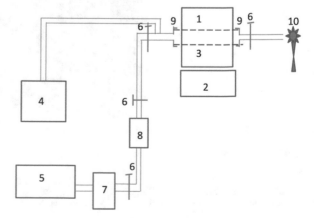

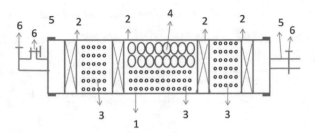

Fig. 4.3 Method of sample positioning in an experimental unit for studying brex behavior when heated in a reducing atmosphere. 1—hollow corundum tube, 2—perforated unit, 3—coke breeze, 4—samples (ore or brex), 5—gate valve, 6—isolation valve

Table 4.2 Physical and mechanical properties of brex after heating at different temperatures

Properties/T (°C)	1000	1100	1200	1300
Porosity (%)	43.48	39.10	32.22	15.02
Density (kg/cm^3)	2.31	2.49	2.82	3.28
Compressive strength (kgF/cm^2)	34	43	47	61

Table 4.3 Main phases of brex from primary oxidized concentrate of the Zhairem plant

Temperature (°C)	Phases
20	Hausmannite (Mn_3O_4); Bixbyite ((Mn, Fe)$_2O_3$); Rhodonite ($MnSiO_3$); Bixbyite C ($FeMnO_3$)
1000	Jacobsite ($MnFe_2O_4$); $Fe_{0.798}Mn_{0.202}O$
1100	$Fe_{0.798}Mn_{0.202}O$
1200	$Fe_{0.798}Mn_{0.202}O$; $Fe_{0.664}Mn_{0.336}O$
1300	$Fe_{2.08}Mn_{0.92}O_4$; $Fe_{0.798}Mn_{0.202}O$
1400	Manganese

It can be seen that iron-manganese melt was formed in the presence of iron oxides. Manganese was subsequently recovered from the melt.

4.2 Brex from Manganese Ore Concentrate and Gas Cleaning and Aspiration Dusts

Chiatura manganese ore concentrate (Republic of Georgia) and baghouse dust from silicomanganese production were used as raw material for brex production. The chemical composition of the manganese ore concentrate and silicomanganese production baghouse dust is given in Table 4.4. The main minerals of the

Table 4.4 Chemical composition of manganese ore concentrate and baghouse dust

Material	Mn	MnO	Fe₂O₃	FeO	Al₂O₃	SiO₂	CaO	MgO	P
Concentrate	35.0–45.0	–	1.2–2.2	–	2.0–3.5	14.0–25.0	2.0–3.5	0.86	0.20
Baghouse dust	–	29.74–31.41	–	0.30	0.01	30.0–34.0	1.67–2.42	1.50	–

concentrate are pyrolusite, psilomelan, and manganite. Pyrolusite from Chiatura manganese ore has a finely-spherulitic (oolitic) and clastic brecciate structure. Spherulites fragments do not exceed 0.25 mm in size. Pyrolusite is partially recrystallized up to hundredths of a millimeter during grain coarsening. The Chiatura oxide concentrate softening point is 1049 °C according to the data in [7]. 85% of baghouse dust particles are smaller than 10 μm. Primary particles have a spherical shape due to phase transitions during particle formation.

Two types of mixtures from the above materials were prepared for brex making: 1—a mixture of an equal mass (50:50) of concentrate and dust, and 2—a mixture of 70% concentrate and 30% dust. The granulometric composition of the second mixture is shown in Fig. 4.4.

Coke breeze was used as a reducing agent; it was provided by the company by whom ferroalloy smelter is owned. The coke breeze has the following composition ratio: carbon—68.72%; volatiles—7.53%; ash—23.75%. 88.46% of coke breeze particles are smaller than 0.635 mm in size. Its moisture content is 9.86%. Three brex mixtures with low binder content were tested in the study (Table 4.5).

Test samples of brex were produced using a computerized laboratory extruder that simulates the processing of a briquetted mixture in commercial even feeders, pug mills and the extruder itself. The brex had a circular cross section of 2.5 mm in

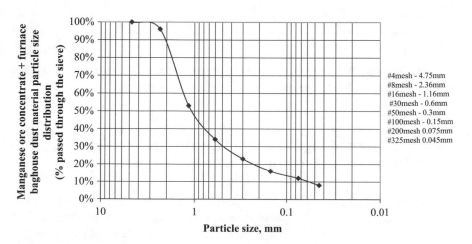

Fig. 4.4 Granulometric composition of mixture for brex production

Table 4.5 Component-by-component composition of brex

Component	No. 1	No. 2	No. 3
Manganese ore concentrate	47.6	66.7	56.0
Coke breeze	–	–	15.0
Baghouse dust	47.6	28.6	24.0
Portland cement	4.8	4.7	5.0

diameter and 1.5–2.0 mm in length. The average moisture content of freshly formed brex was 11%. The vacuum level in the working chamber of the extruder was maintained at 38–48 mm Hg.

Due to the cylindrical shape of brex, its cold strength was determined by measuring both compressive strength and tensile splitting strength, since these types of stress would be exerted on brex during actual transport and stacking. These properties were measured using an Instron 3345 system (USA). The compressive and tensile splitting strength tests were performed using six specially-prepared brex samples with a diameter of 25 mm and a length of 20 mm which were subjected to compression and tensile splitting.

Figures 4.5 and 4.6 illustrate the behavior of brex and the mechanism of brex destruction during compressive and splitting strength tests. At the initial stage (1), the elastic loading of brex shows a linear dependence between the applied force and the displacement of the active pressing surface. The beginning of the second stage is characterized by a sharp decrease in the applied force due to the formation of surface fractures. At this stage, the "near-surface" layers of the brex are destroyed and peeled (15–20% of the radius). However, the "core" (central part) of the brex

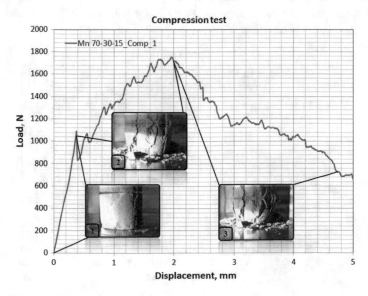

Fig. 4.5 Compression test of brex No. 3

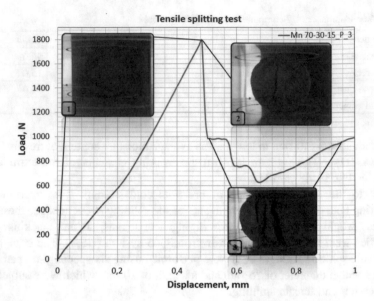

Fig. 4.6 Tensile splitting test of brex No. 3

continues to endure the pressure, as evidenced by the core's growth. A clear zonal structure of the brex can be observed in the radial direction. The near-surface layers of the brex have a relatively higher stiffness than its "core". This statement is supported by the difference in slopes of the curves at Stage 1 and Stage 2.

In the third stage, once the applied force reaches its maximum, the "core" of brex is destroyed with a complete strength loss.

Brex No.1 and Brex No.3 demonstrated similar behavior (Figs. 4.7 and 4.8).

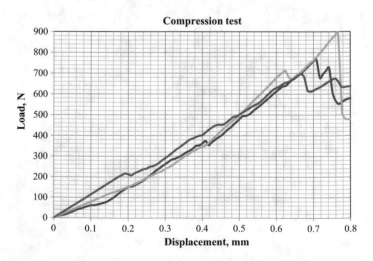

Fig. 4.7 Compression test of brex No. 2 (for three different samples of the brex No. 2)

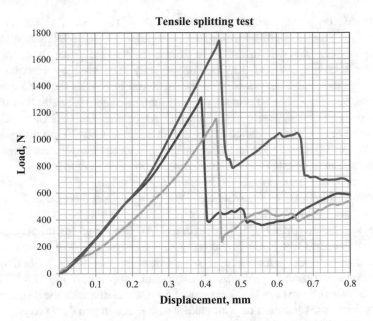

Fig. 4.8 Tensile splitting test of brex No. 2 (for three different samples of the brex No. 2)

Table 4.6 illustrates the average compressive strength and splitting strength.

Subsequently, compressive strength of samples of industrially-manufactured brex No. 2 was tested by independent laboratory L. Robert Kimball & Associates, Inc (USA). The compressive strength of a sample with a diameter of 32.3 mm and a length of 57.7 mm was 206 kgF/cm^2. It is clear that the results obtained for compressive strength significantly exceed the compressive strength of briquettes for SEAF. The above-mentioned results for brex compressive strength were achieved by using a relatively low Portland cement content compared with that of vibro-pressed briquettes (8–9% of weight). Most of a batch of brex manufactured for pilot testing contained only 3% of a binder.

The brex porosity was examined by a LEO 1450 VP scanning electron microscope (Carl Zeiss, Germany) with a resolution of 3.5 nm in combination with X-ray computer tomography using the Phoenix V|tome|X S 240 high-resolution micro focus computed tomography system (General Electric, USA). Computer tomography determined that the sample porosity was in the range of pores above 100 μm in diameter. Scanning electronic microscopy was used to study the porosity

Table 4.6 Mechanical strength of brex (kgF/cm2)

Brex No.	Tensile splitting strength	Compressive strength
1	18.3	160.8
2	28.6	291.2
3	13.1	124.6

represented by pores with dimensions less than 100 μm. The results of the microscopic studies were processed by computer program STIMAN [8]. The porosity of brex was determined to be 18.5% (5.6%—the porosity, measured by computer tomography, and 12.9%—the porosity measured by electron microscopy). Subsequently, the porosity of manufactured brex with the same content of basic components (3% of Portland cement) amounted to 18.7%, as measured by independent laboratory L. Robert Kimball & Associates, Inc (USA).

Thermal analysis of powdered samples with a mass of 50–70 mg using the STA 449 C Jupiter® (Germany) in an argon atmosphere, in the temperature range of 20–1400 °C at a heating rate of 20 °C/min, made it possible to determine the character of the brex phase transformations. In addition, the results of the analysis showed that the thermal effects on the DSC curve for brex No. 1 and brex No. 2 are almost identical (see Fig. 4.9).

The manganite (MnO (OH)) dehydrates to form pyrolusite or β-kurnakit ($2MnOOH = Mn_2O_3 + H_2O$) in the temperature range 300–450 °C. At temperatures above 400 °C, psilomelane of manganese ore transforms to hollandite or hausmannite. Exothermic peaks with a maximum temperature of 795.8 °C in brex No. 1 and 801.0 °C in brex No. 2 are most likely associated with the dissociation of pyrolusite and the formation of β-kurnakit which is accompanied by oxygen release and mass loss. The same peak in brex No. 3 at a temperature of 742.8 °C is not followed by a loss of mass; it can be explained by the effects of recrystallization of amorphous phases. β-kurnakit decomposes in the temperature range 900–1050 °C, and the formation of β-hausmannite takes place [9]. In the range 1080–1250 °C, β-hausmannite is polymorphically converted to γ-hausmannite.

In brex No. 3, the endothermic peak in 150–200 °C temperature range is accompanied by mass loss due to sorption moisture removal. In 300–500 °C temperature range there are no pronounced endothermic peaks, but there is a loss of mass associated with the reduction of MnO_2 to Mn_2O_3. A phase transformation of manganite (MnOOH) into kurnakit (Mn_2O_3) occurs in the same temperature range. The endothermic effect with a maximum at 1031.7 °C is associated with the reduction of Mn_3O_4 to MnO due to mass loss; the effect at 1056.1 °C is determined by the ferric phase change. Mössbauer analysis was performed to determine the phase composition of the iron-containing components in the original brex and the products of its heat treatment, and to clarify the results of the thermogram. Figure 4.10 shows the Mössbauer spectra of raw brex and the samples after heating at different temperatures. Table 4.7 illustrates the results obtained after processing the spectra using the computer program Univem.

The Mössbauer spectrum of the initial brex sample (Fig. 4.10a) is a superposition of two doublets, with the isomeric shift of the doublets being smaller than that of iron oxides and iron hydroxides. These doublets can be attributed to the intermetallic iron and manganese compound. Due to the isomeric shift being close to 0 and a small quadrupole splitting, the doublet D2 can be attributed to the iron atoms localized in the cubic face-centered lattice. These iron atoms do not have manganese atoms in their environment and the doublet D1 is attributed to iron ions with

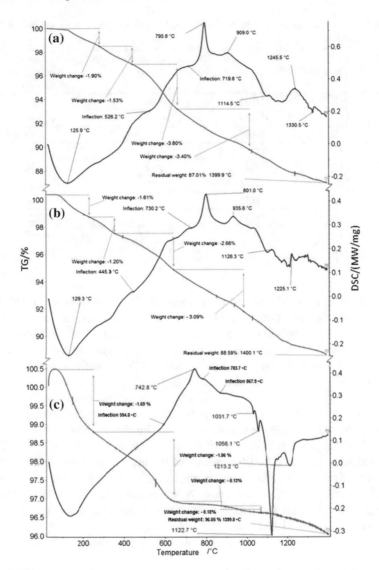

Fig. 4.9 Thermograms of the brex samples No. 1 (**a**), No. 2 (**b**), No. 3 (**c**) [left hand vertical—TG/ %; right hand vertical—DSC/(MWt/mg)]

manganese atoms present in their coordination sphere. As a result, the quadrupole splitting of this doublet D1 is notably higher.

The Mössbauer spectrum of the initial brex sample heated to 850 °C and cooled in the STA conditions (Fig. 4.10b) has shown no significant difference from the initial sample. There was practically no change in the isomeric shift and the quadrupole splitting of doublets (see Table 4.7). However, the ratio of the doublet areas has changed; this implies that there could be only a small change in the

Table 4.7 Mössbauer parameters of raw brex No. 3 and brex No. 3 after thermal analysis at various temperatures

Sample	Spectrum component	Isomeric shift δ (mm/s)	Quadrupole splitting Δ (mm/s)	Area component, S (%)
Raw brex	D1	0.22	0.56	74.04
	D2	0.07	0.30	25.96
850 °C	D1	0.23	0.56	84.83
	D2	0.07	0.27	15.17
1065 °C	D1	0.17	0.59	52.98
	D2	0.11	0.24	47.02
1175 °C	D1	0.17	0.65	46.80
	D2	0.13	0.22	53.20
1250 °C	D1	0.19	0.61	48.83
	D2	0.13	0.21	51.17

positional distribution of iron and manganese atoms in the sample. Therefore, the thermal effects described on the DSC curve are not related to the iron-containing phase, but to the manganese phase. The changes in the spectra occurred due to thermal effects at 1031.7 and 1056.1 °C, which is reflected in the spectrum after heating up to a temperature of 1065 °C under STA conditions. Meanwhile, the quadrupole splitting of the doublet D1 and the doublet area decreases, but the corresponding parameters of the doublet D2 increases; this can be attributed to the sample phase transformation and change in the density of structural positions. After phase transformations at these temperatures, no significant changes occurred in the spectra. Consequently, the thermal effects observed at 1122.7 and 1213.2 °C are associated with the phase transformations of the manganese component in brex.

The endothermic peak with a maximum at 1122.7 °C is determined by forming manganese carbide Mn_3C, where carbon recovers manganese from MnO. The coke breeze in brex provides for intensive iron and manganese reduction, marked by endothermic peaks at 1122.7 and 1213.2 °C.

The brex structure has been studied by electron microscopy. Figure 4.11 illustrates the structure of brex No. 2.

Portland cement is known to lose its binding properties at 750–900 °C. Therefore, it is important to understand what provides for the hot strength of brex at temperatures higher than the above-mentioned temperature range. A previous studies [10] explored the behavior of a pyrolusite sample of manganese concentrate from the Chiatura mine when heated in a reducing atmosphere (helium 40% and hydrogen 60%; heat at a rate of 20 °C per minute). A cubic sample measuring $5 \times 5 \times 5$ mm was observed to gradually expand by a considerable amount up to 0.8 mm; the thermal expansion was accompanied by sample cracking. Cracks were first observed at 500 °C, with the number of cracks increasing with heating. The fractures formed during the tests are characterized by a certain orientation pattern.

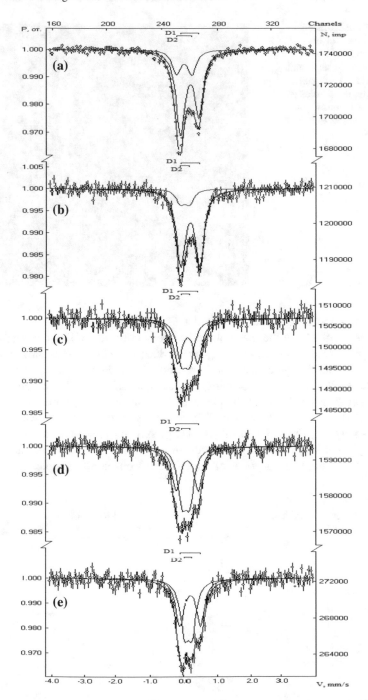

Fig. 4.10 Mössbauer spectra of raw brex No. 3 and of brex No. 3 after heated to STA conditions to temperatures: **a** crude; **b** 850 °C; **c** 1065 °C; **d** 1175 °C; **e** 1250 °C

Fig. 4.11 Backscattered
electron image of brex
No. 2 structure. 1—
pyrolusite, 2—gaussmanite,
3—diopside, 4—fayalite,
5—mullite, 6—pyrolusite,
7—solid solution of
Ca_2SiO_4–Mn_2SiO_4

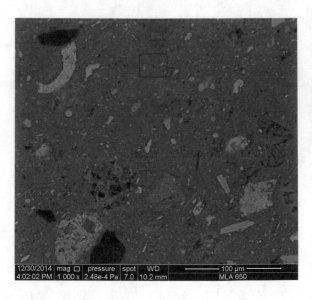

Radial and concentric cracks are found within the spherulites. Cracks in random directions are found in the inter-spherulites cementing mass. Despite this fracture pattern, the sample maintained integrity at 800 °C. After that, a slow and irreversible compression process began. In our opinion, a dense silicate phase (olivine or Wollastonite) was the cementing binder that safeguarded the sample from destruction. The formation of the phase in the indicated temperature range corresponds to the well-known phase diagrams of systems Ca_2SiO_4–Mn_2SiO_4 and $CaSiO_3$–Mn_2SiO_3 (Figs. 4.12 and 4.13 [11]). A glass phase and manganese silicates (tephroite-Mn_2SiO_4 and rhodonite-$MnSiO_3$) were discovered in the intervals between the oxide phase's grains of the Chiatura concentrate sample.

Fig. 4.12 Diagram of system
Ca_2SiO_4–Mn_2SiO_4

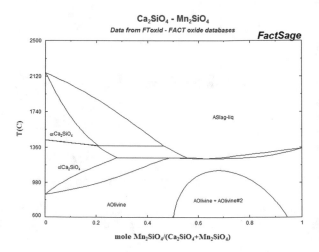

Fig. 4.13 Diagram of system
CaSiO₃–MnSiO₃

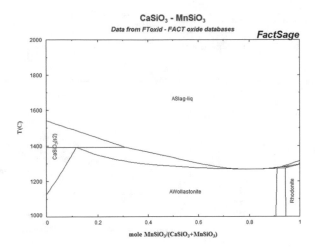

The dense structure of the above-mentioned silicate phases provides the strength of brex at a temperature up to 1250 °C. At a higher temperature, two eutectics are formed in the MnO-SiO₂ system: (1) tephroite + rhodonite + liquid (1251 °C) and (2) tephroite + manganosite + liquid (1315 °C). These eutectics contribute to brex strength retention (see Fig. 4.14) [12], due to the presence of the liquid binder.

An important property of brex for smelting ferroalloys in SEAF is the specific electric resistance. The measured specific electrical resistance of brex samples at a room temperature is given in Table 4.8.

The measuring system used in the study made it possible to investigate the dynamic pattern of brex resistance at a temperature up to 800 °C. As a result, the behavior of brex resistance was studied up to the first exothermic peak described in Fig. 4.9. It corresponds to a relatively deep brex immersion in the furnace to levels where lower manganese oxide is formed. Due to the low electric resistance of brex No. 3 (with 15% of coke breeze), which is lower than the resistance of lump manganese ore of the same deposit [13], this type of brex was excluded from further study.

Fig. 4.14 Phase diagram of
the MnO–SiO₂ system

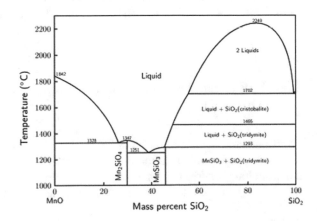

Brex, No.	ρ, Mohm mm	
1	26.3	25.5
2	32.3	28.9
3	9.42	12.4

Table 4.8 Specific electric resistance of brex No. 1– No. 3 at room temperature

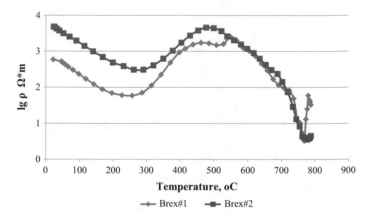

Fig. 4.15 Specific electric resistances of brex No. 1 and brex No. 2

Figure 4.15 shows the results of resistance measurements in brex No. 1 and brex No. 2 when heated to 800 °C.

The increase in resistivity of brex in 300–500 °C temperature range coincides with the above-mentioned endothermic effects of mass loss, which is associated with manganite dehydration and with kurnakit formation.

It is clear that the resistivity of brex No. 1 is lower than that of brex No. 2 because of the higher carbon content in brex No. 1 resulting from a higher baghouse dust content that adds carbon to brex No. 1.

Based on the above-mentioned conclusions regarding the physical and mechanical properties of brex, the behavior of brex during heating, and measurements of the specific electrical resistivity, brex No. 2 was selected as a component of the SEAF charge.

4.3 Full-Scale Testing of Silicomanganese Smelting with Brex

To manufacture a brex batch lot of 2000 tons a brickmaking plant equipped with a J. C. Steele industrial extrusion line was selected in Ragland (Alabama, USA). One thousand four hundred tons of the Manganese ore fines from the Republic of Georgia (Chiatura) and 600 tons of the baghouse aspiration dusts from the Client's ferroalloys plant in West Virginia were delivered for brex production. The production process consisted of blending the dust and ore volumetrically via a front-end loader into blending stockpiles. Because the ore came in coarser than desirable, the material blend was fed into the plant through the grinding facility where the material was reduced to minus 8 mesh (2.25 mm). The blended and ground material was transferred into the feed tanks in the extrusion plant. The blend was further fed into Steele 75ADC Extruder where it was mixed with water and 3–5% of PC. The extruder and its vacuum mixer had a combined 338 KW capacity. The Extruder was equipped with multi-hole extrusion die with round openings of 25 and 30 mm diameter. Typical production parameters were as follows: production rate 55 metric tons per hour; moisture content of the green brex 10.5%; vacuum level: 100 mm Hg. The appearance of the brex on the outlet of the extruder and in the storage area after just 5–15 min after production is shown in Fig. 4.16. As expected, the green brex handling procedure did not require any special protective measures or special treatment (steam curing, palletizing to preserve the integrity, etc.). The brex were shipped to the smelter by barges discharged at the berth of the Ohio River (Fig. 4.17) Trucks were then used to transport the brex from the pier to the open batch stockyard a conveyor then carried them directly to the furnace bunkers. In total, the transport of the brex to the ferroalloy (2000 km) withstood 20 handling operations in a period of 30 days: manufacturer—extruder, conveyors, dumps truck, stockpiles, wheel loader, truck; port of loading—stockpiles, wheel loader, hopper, bucket, barge; port of discharge—grab, hopper, conveyors, truck; ferroalloy plant ore storage yard—stockpiles, front loader, factory warehouse, furnace. Total fines (less than 6 mm) generated during these operations did not exceed 10%.

To initiate a full-scale industrial trial, a stable 27MVA capacity and 85-tons/day-average industrial submerged EAF was selected to run with a specific average energy consumption of 4200 kWh/t, manganese recovery rate of 80%, and manganese dump slag content of 12–14%. The chemical composition of manganese-containing charge components is shown in Table 4.9.

Fig. 4.16 Production of the brex by Steele 75 Extruder (left) and discharge of the green brex after 5–15 min at the storage area (right)

Fig. 4.17 Process of
unloading brex from barge

Table 4.9 Chemical composition of charge components

Material	Components composition						
	Mn	SiO$_2$	CaO	MgO	Al$_2$O$_3$	Fe	P
Ore-1	49.5	13.0	0.7	0.5	1.0	4.0	0.05
Ore-2	29.0	20.2	5.9	5.2	2.1	0.9	0.06
Brex	31.37	24.32	6.10	2.05	2.79	1.36	0.12
Scraps	23.3–38.6	N/A	N/A	N/A	N/A	N/A	N/A
Silicomanganese fines briquettes (vibropressed)	53	18	4	N/A	N/A	10.5	0.15

Scraps are the internal wastes generated while deslagging the surface of metal in a ladle, cleaning buckets and furnace notch, sorting at slag dumps, and casting Silicomanganese. Only the total manganese content in slags was determined; these values are given in Table 4.10. Briquettes made from Silicomanganese fines are produced through the process of vibropressing (unconditioned fines fraction 0–6 mm). To impart the necessary strength to the briquettes made from these fines, significantly more Portland cement was required than is for brex (10%).

Table 4.10 Composition of charge (net ton)

Brex share (%)	Basic elements content (% mass)								
	In metal				In slag				
	Mn	Si	Fe	P	MnO	MgO	SiO$_2$	CaO	Al$_2$O$_3$
0	66.64	17.71	14.06	0.127	10.50	7.38	44.28	21.43	15.29
5	66.33	16.30	14.25	0.128	12.70	7.14	45.33	20.31	13.46
10	66.90	17.69	14.18	0.098	12.89	6.52	44.78	18.67	13.80
20	66.77	16.45	14.21	0.12	11.30	6.77	44.57	19.86	15.07
29	68.02	16.02	14.01	0.14	11.79	6.39	45.11	19.42	15.13
35	65.98	15.22	14.30	0.16	15.08	5.73	46.20	26.11	15.71
40	66.08	17.07	14.25	0.18	10.8	5.17	45.30	19.83	15.35

A preliminary calculation of the charge composition was made in order to achieve stable manganese content in the ore part of the charge (Table 4.11). With the increase in the share of brex in the ore part of the charge, it was decided to reduce the supply of ore-2, which is a drop-in replacement, because the average manganese content of brex was 31% against 29% in the said grade.

For the accuracy of the comparison of the results of the furnace operation with and without brex, a month-long period of furnace operation was observed.

Table 4.11 Composition of the charge (net tons)

Component of the charge	Reference and full-scale trial periods						
	Reference period	1	2	3	4	5	6
Mn Ore-1	0.526 (30%)	0.525 (30%)	0.525 (30%)	0.525 (30%)	0.525 (30%)	0.525 (30%)	0.525 (30%)
Mn Ore-2	1.205 (70%)	1.120 (65%)	1.030 (60%)	0.855 (50%)	0.705 (41%)	0.600 (35%)	0.525 (30%)
Brex	–	0.087 **(5%)**	0.175 **(10%)**	0.350 **(20%)**	0.500 **(29%)**	0.605 **(35%)**	0.690 **(40%)**
Estimated weight of charge	1.730	1.732	1.730	1.730	1.730	1.730	1.740
Incoming manganese with:							
Ore-1	0.262 (43%)	0.262 (42.6%)	0.262 (42.6%)	0.262 (42.4%)	0.262 (42.2%)	0.262 (42.1%)	0.262 (41.8%)
Ore-2	0.350 (57%)	0.325 (53%)	0.299 (48.6%)	0.224 (40.1%)	0.205 (32.8%)	0.174 (27.9%)	0.152 (24.2%)
Brex	–	0.027 (4.4%)	0.054 (8.8%)	0.109 (17.5%)	0.160 (25%)	0.188 (30%)	0.214 (34%)
Manganese charge weight	0.612	0.614	0.614	0.619	0.622	0.624	0.628
Average manganese content (%)	35.4	35.5	35.6	35.8	36.0	36.1	36.1

A weeklong period of the furnace operating without brex immediately preceding the beginning of the pilot period was taken as a reference.

We decided to begin with 5% of the brex share in the ore part of the charge to get a first experience with the agglomerated burden. Over the course of 3 days of observation, no visible changes had been registered in the process of Silicomanganese smelting; the furnace worked smoothly, with a constant and uniform current load during the period of brex use. The slight 0.3% decrease in manganese extraction during this period was due to furnace downtime associated with the tap-hole unit repair (1.5 h downtime). Then, for four consecutive days, the brex share in the charge was maintained at 10%. Improvement in the furnace top functioning was visually apparent. Gas flames throughout the furnace's top area confirmed an improvement in gas permeability of the column of the charge and the uniformity of the temperature distribution over the surface of the furnace top. Deeply submerged electrodes functioned without producing surface blowholes in the electrodes circle and in the surrounding area. No sintering in the electrodes circle was observed. During the next 3 days, brex share in the charge was increased up to 20%. The furnace operated smoothly, the electrodes were deeply submerged, the current load had no visible abnormalities or tremors, and the gas permeability of the charge was good for the entire area of the furnace top. Throughout the next week, the brex share of the charge was increased to 29%. The furnace worked well. An increase in the rate of the charge descent in the electrodes circle, especially during the tapping, indicates an increase in the melting rate of the charge in the active zone of the furnace. The furnace top functioned smoothly without surface blowholes with a constant load current, smooth, without bumps and drops. Melt out was good, metal and slag were sufficiently warmed up. However, in the end of this period, there was a furnace downtime for 4 h due to reasons not related with the brex presence in the charge (electrical problem in the transformer).

The beginning of the next phase of the pilot period, when the brex share rose to 35% substituting the ore-2, was associated with a problem caused by a disruption in the tap-hole electrode bypass, resulting in its shortening and the deterioration of the melt. The bulk of the slag remained in the furnace. This resulted in an increased coke rate and scraps withdrawal from the charge along with briquettes of Silicomanganese fines. During this period, a deterioration in the technical and economic performance of the furnace had been registered, namely an increase in specific energy consumption and specific consumption of manganese ore raw materials, as well as a reduced extraction of manganese. To restore its normal functioning, the furnace heating rate was increased so as to warm up the slag and ensure its normal tapping. In the final phase (week) of the pilot operation period, the furnace worked with the brex share in the charge up to 40%. The furnace operation was characterized by a good current load, top gas release was smooth, with no surface blowholes and furnace charge downslides, and with a normal tapping.

Based on the results of the full-scale trial, we made a comparative analysis of the technical and economic performance of the furnace with and without brex in the charge. Technical and economic performance of the furnace and the chemical

Table 4.12 Furnace parameters during the reference and the full-scale trial period

Parameter		Reference and full-scale trial periods						
		Reference period	1	2	3	4	5	6
Actual metal production over a period of time (t)	Ton	816.323	298.4	277.1	196.6	570.6	397.2	757.2
	b.t (Basic ton)	839.670	300.7	285.8	199.5	584.75	393.2	767.8
Actual furnace performance (%)		98.9	97.6	97.4	99.7	96.2	98.6	93.6
Power consumption (MW)		3,339.27	1078.2	1085.4	734.7	2120.8	1536.3	2821.48
Specific power consumption (kW*h/b.t)		3977	3586	3798	3682	3627	3908	3675
Ore-2	t(29% Mn)/b.t	1.106	0.933	0.892	0.714	0.635	0.626	0.492
	t(48% Mn)/b.t	0.668	0.563	0.539	0.431	0.383	0.378	0.297
Ore-1	t(49.5% Mn)/b.t	0.565	0.505	0.482	0.509	0.480	0.524	0.484
	t(48% Mn)/b.t	0.582	0.520	0.497	0.525	0.495	0.540	0.499
Brex	t(31.37% Mn)/b.t	0	0.077	0.164	0.273	0.387	0.571	0.605
	t(48% Mn)/b.t	0	0.050	0.107	0.178	0.252	0.373	0.395
The total consumption of raw manganese ore	t/b.t	1.671	1.515	1.538	1.496	1.502	1.721	1.581
	t(48% Mn)/b.t	1.250	1.133	1.143	1.134	1.130	1.291	1.191
Coke (t/b.t)		0.446	0.339	0.420	0.405	0.395	0.415	0.413
Quartzite (t/b.t)		0.419	0.499	0.524	0.465	0.529	0.456	0.475
Silicomanganese fines Briquettes (t/b.t)		0.158	0.082	0.103	0.092	0.120	0.110	0.094
Scrap (Mn content in scrap, %) (t/b.t)		0.358 (23.3)	0.601 (29.9)	0.462 (33.0)	0.477 (35.3)	0.461 (32.0)	0.455 (25.8)	0.373 (38.6)
Electrode mass (t/b.t)		0.034	0.032	0.030	0.035	0.028	0.033	0.028
Manganese extraction from the ore component (%)		80.1	79.8	80.7	80.7	83.6	79.1	79.9

analysis of the fusion products are presented in Tables 4.12 and 4.13. All the numbers are taken from the foundry journals recordings.

The main positive aspect of the full-scale campaign with the brex in the charge for the smelting of merchandise Silicomanganese is that the furnace operated in a stable, smooth regime. Furnace top operation was characterized by good gas permeability all over the surface, without any charge downslide. Current load was uniformly distributed among three electrodes. Electrodes were submerged deep and

Table 4.13 Main components composition during the reference and pilot operation period

Phase	Main components composition								
	In metal				In furnace charge				
	Mn	Si	Fe	P	MnO	MgO	SiO$_2$	CaO	Al$_2$O$_3$
Reference	66.64	17.71	14.06	0.127	10.50	7.38	44.28	21.43	15.29
1	66.33	16.30	14.25	0.128	12.70	7.14	45.33	20.31	13.46
2	66.90	17.69	14.18	0.098	12.89	6.52	44.78	18.67	13.80
3	66.77	16.45	14.21	0.12	11.30	6.77	44.57	19.86	15.07
4	68.02	16.02	14.01	0.14	11.79	6.39	45.11	19.42	15.13
5	65.98	15.22	14.30	0.16	15.08	5.73	46.20	26.11	15.71
6	66.08	17.07	14.25	0.18	10.8	5.17	45.30	19.83	15.35

stably. Melt tapping took place according to the schedule, the chemical composition of the metal and slag showed no significant changes. Replacement of the substantial part of the lumpy manganese ore by brex based on ore fines and aspiration dust led to the improvement of technical and economic indicators of the process as a whole. Specific energy consumption during the test period decreased significantly. In the reference period, consumption per basic ton of the alloy was 3977 kWh with the brex share in the charge ore equal to 40%; the specific energy consumption decreased up to 3675 kWh per basic ton (−7.6%).

Another positive result of the full-scale trial is related to increased manganese extraction from the ore. In 29% of brex in the charge, manganese extraction was at 83.6%, against an average extraction of 80% in the reference period of the furnace operation without brex in the charge (Fig. 4.18). Decreased extraction in the period preceding the final phase was not associated with the presence of brex in charge, but was, instead, the result of furnace downtime and problems with the electrode.

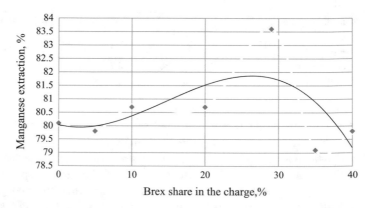

Fig. 4.18 Manganese extraction as a function of the brex share in the ore component in the furnace charge

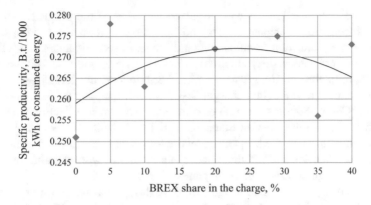

Fig. 4.19 Specific productivity, B.t./1000 kWh of consumed energy

It is also important to know the relationship of the brex share in the ore part of the charge and the specific productivity of the furnace expressed in base tons per unit of electricity consumed. Figure 4.19 shows that the best performance is achieved when the brex share in the ore part of the charge is in 20–30% range. In general, the results of full-scale trials give grounds to consider that it is appropriate to use brex based of manganese ore fines and baghouse dust from gas cleaning as the essential charge component for Silicomanganese smelting.

4.4 Stiff Extrusion Briquetting of the Ferroalloys Fines

Before shipping to consumers, ferroalloys are crushed and screened to obtain a required piece size. During such processing ferroalloys fines and aspiration dust are generated. One of the ways to utilize the ferroalloy fines and the dusts is briquetting. The basic requirements for quality of commercial briquettes made from fines of ferroalloys are reduced to the necessity to provide the required mechanical strength at lowest possible binder content and high thermal stability, i.e. the ability to maintain integrity when placed in the environment with high temperature. Briquetted ferroalloys used as components of the SEAF charge must have sufficiently high values of hot strength and of specific electrical resistance. In order to increase the specific electrical resistance, dust of aspiration systems can be added to metallized ferroalloys fines. In addition, aspiration dusts of ferroalloys production, as shown above, can exhibit binding properties which can significantly reduce binder consumption.

The field experience in using roller presses for agglomeration of manganese-containing, and hence abrasive ferroalloys fines, confirmed that this method has a number of disadvantages. In August 2003, Nikopol Ferroalloys Plant (Ukraine) launched the experimental production of ferroalloys-fines-based briquettes using an organic binder. The capacity of the briquetting line allowed producing 40,000 tons of briquettes per year [14]. It was projected that 70% of the briquettes would be produced from silicomanganese fines and 30% from ferromanganese fines. In the process of the production of silicomanganese-fines-based briquettes, the following parameters were achieved: productivity, t/h—4.5 ... 5.0; briquette volume, cm³— 22 ... 25; briquette density, g/cm³—4.5 ... 4.8; briquette moisture, %—0.8 ... 1.7; moisture content in briquetted charge, %—2.5 ... 4; temperature of charge before pressing, °C—60–80; drop strength of briquettes (+5 mm size fraction when tested in accordance with the Russian Standard GOST 25471-82), %—94 ... 96.

When implementing the briquetting technology, an increased wear of the sleeves surface was noted. As a result, the first set of sleeves was worn out after briquetting 1100 tons of high abrasive silicomanganese fines.

The Assmang plant in Cato Ridge (South Africa) had a successful experience in using vibropress briquetting technology to agglomerate manganese-containing anthropogenic materials, including ferroalloys fines [15]. The initial composition of the tested briquettes consisted of 69.2% ferromanganese fines, 26.8% of aspiration dust and 4% of cement binder. Mechanical strength was tested using a drop test procedure, which showed that less than 10% of fines with smaller than 0.5 mm size particles are formed. However, the tests that were carried out in a muffle furnace with heating in the range of 200–800 °C showed that the briquettes of the above-mentioned composition do not have hot strength. As the temperature increased with the interval of 200 °C, the briquettes were removed from the furnace and subjected to a strength test. It was noted that the briquettes completely lost their strength at 400 °C. At the next stage briquettes with alumina cement as a binder were tested, which allowed achieving specified hot strength. However, the mechanical strength of this type of briquettes did not meet the requirements. In addition, the use of alumina cement would lead to a significant increase in the briquette cost. The mixture of cement and clay rich in alumina (3% by weight) was decided to use as a binder, which eventually resulted in the production of briquettes with the sufficient strength for a 12-h long stacking and with acceptable mechanical strength after 48 h of storing. In addition, the briquettes had sufficient hot strength. Since the cost of clay is 50% lower than the cost of alumina cement, the final composition of the briquettes is the following: 67.0% ferromanganese fines, 26.0% aspiration dust, 4% cement binder and 3% aluminum-rich clay. The chemical composition of the briquette is given in Table 4.14.

Table 4.14 Chemical composition of Assmang briquette

Elements	Content (%)
Mn	56.2
Fe	9.84
SiO_2	4.46
CaO	3.63
MgO	0.58
C	6.28
Al_2O_3	1.83
Mn/Fe	6/1

After that, the briquettes were used as charge components for SEAF (11.5 MVA). The briquettes replaced two grades of manganese ore, namely Nchwaning and Gloria. The share of briquettes in the furnace charge was 10%. During this full-scale testing, no rise of electrodes was observed; no charge sintering occurred. The use of ferromanganese fines-based briquettes as charge components resulted in an improvement in the technical and economic parameters of the furnace operation.

In order to study possible ways of using the SVE technology for agglomeration of similar materials, the following types of brex were tested: brex from metallized crushing fines and brex with the addition of aspiration dusts of the ferroalloys crushing units.

The briquettes were manufactured on a laboratory extruder with the hole size of 15 mm and the cylindrical shape of brex. The vacuum was 37.5–62.5 mmHg. The brex composition and the physical and mechanical properties of brex are given in Table 4.15. The mixture for Brex No. 4 was subjected to a 24-h homogenization. Figure 4.20 illustrates silicomanganese crushing fines and brex based on them.

Table 4.15 Composition of brex based on silicomanganese fines

Composition of Brex	No. 1	No. 2	No. 3	No. 4[a]
Silicomanganese fines (%)	94.0	93.0	92.0	94.0
Portland cement (%)	5.0	5.5	6.0	5.0
Bentonite (%)	1.0	1.5	2.0	1.0
Breaking load for brex (after 24 h) (kgF)	18.4	29.4	32.8	20.9
Fines formed during the drop test after 24 h (less than 4.75 mm) (%)	6.9	3.8	3.3	8.9
Failure load on brex (after 7 days) (kgF)	62.4	66.3	79.9	64.7
Fines formed during the drop test after 7 days (less than 4.75 mm) (%)	3.5	2.5	3.1	3.0
Density (g/cm³)	4.13	4.10	4.08	4.02

[a]With homogenization during 24 h

Fig. 4.20 Silicomanganese crushing fines (left) and brex based on Silicomanganese crushing fines (right)

It is evident that the brex achieved sufficiently high compressive strength and impact strength, and homogenization did not have an obvious effect. The increase in cement content from 5 to 6% and in bentonite content from 1 to 2% improved compressive strength by 28% and impact strength by 12%. Applicable strength level was achieved with a minimum content of cement (5%) and bentonite (1%).

The composition of brex based on ferroalloy fines with addition of the aspiration dust collected at the ferroalloys crushing unit was tested as well: (1) silicomanganese fines—80.2%, aspiration dust of silicomanganese production—14.2%, Portland cement—4.7%, bentonite—0.9%; (2) silicomanganese fines—66.1%, aspiration dust of silicomanganese production—28.3%, Portland cement—4.7%, bentonite—0.9%).

The crushing strength of these brex was, on average, according to the results of measurements on 10 samples 244.6 kgF/cm^2 and of 116.8 kg/cm^2, respectively.

The dynamics of the electrical conductivity of the brex from the first batch was examined during the heating of the sample till 800 °C (Fig. 4.21).

For this type of brex the room temperature electrical conductivity is 101.2 (Mohm m)-1. For the brex used in the above-described industrial pilot tests of silicomanganese smelting, the electrical conductivity was 207.7 (Mohm m)$^{-1}$ at 20 °C. When heated to a temperature of 750 °C, the conductivity of both types of brex were 4,463,670.5 (Mohm m)$^{-1}$ and 120,328.2 (Mohm m)$^{-1}$, respectively. The

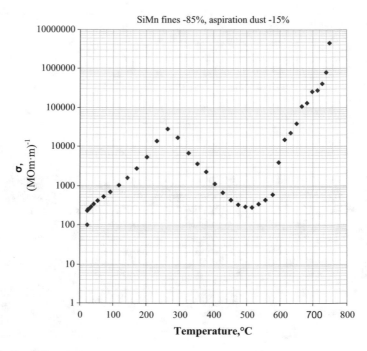

Fig. 4.21 Dynamics of changes in the electrical conductivity of brex based made of manganese fines when heated

Fig. 4.22 Brex tested in
Tamman furnace

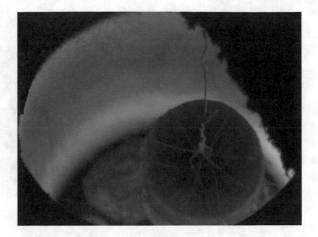

electrical conductivity of the silicomanganese fines-based brex was 37 times higher than that of the brex from manganese oxides. To reduce the electrical conductivity, it is necessary to increase the content of aspiration dust in the brex.

The thermal stability tests of the brex were carried out in a Tamman furnace by holding the brex above the melt for 15 min at a temperature of 1200 °C. All samples passed the test and maintained their integrity. In addition, the brex were also smoothly melted after being dropped into a slag melt at a temperature of 1580 °C. It took 10 min to melt a single brex unit (with a diameter and length of 25 mm) (Fig. 4.22).

In recent years, a number of ferroalloys producers have fostered efforts to develop an efficient technology for briquetting ferrosilicon crushing fines. The paper [16] examines the results using liquid glass (sodium silicate) as a binder to produce strong briquettes from this material. The required strength of the briquettes was achieved by adding liquid glass in the amount of about 6% of the weight of a given briquette batch; however, use of the same binder (2%) in combination with starch (about 3.0%) resulted in strength gain. It has been stated that the results of the study provided the basis for the design and construction of three industrial lines for briquetted ferrosilicon production. The briquetting line of Silingen Polska [17] is known to produce about 18,000 tons of these briquettes per year. The maximum compression strength of such briquettes is 14 MPa. The briquette dimensions are 66 mm × 55 mm × 35 mm and composition is 68% silicon and up to 2% carbon, due to the use of an organic binder, which exceeds the carbon content limit in ferrosilicon. The content of liquid glass-supplied Na_2O is not indicated.

We tested briquettes from ferrosilicon FS 75 grade crushing fines with added aspiration dust from FS 75 crushing units. The exterior appearance of the briquetted material is illustrated in Fig. 4.23. Composite binder comprised of different grades of lignin and starch was used in the testing.

The chemical composition of the aspiration dust is given in Table 4.16. More than 99% of the particles are smaller than 0.125 mm in size.

Fig. 4.23 Crushing fines of ferrosilicon FS 75

Table 4.16 Chemical composition of aspiration dust formed in the crushing sections of ferrosilicon FS 75

Material	Chemical composition							
	Si	Al	Ca	Cr	Mn	C	P	Fe
Aspiration dust of FS 75 crushing unit	78.83	1.49	0.58	0.13	0.21	0.21	0.03	18.52

Three brex mixtures have been developed and tested with 4-h preliminary homogenization and without it. The brex exterior appearance is shown in Fig. 4.24. The component composition of the brex and the extrusion parameters of brex production are presented in Table 4.17.

Fig. 4.24 Brex from FS 75 fines and aspiration dust

Table 4.17 Component composition of brex from FS 75 fines and aspiration dust and extrusion parameters

Sample No.	1	2	3	4	5	6
FS 75 fines (%)	81.0	81.0	81.0	81.0	82.0	82.0
Aspiration dust (%)	9.6	9.6	9.6	9.6	9.8	9.8
Microsilica (%)	4.8	4.8	4.8	4.8	4.8	4.8
Bentonite (%)	1.0	1.0	1.0	1.0	1.0	1.0
Composite binder (%)	3.5	3.5	3.5	3.5	2.5	2.5
Strengthening time (h)	0.0	4.0	0.0	4.0	0.0	4.0
Moisture content during extrusion (%)	5.6	5.9	5.1	5.7	7.1	6.8

Prior to the tests, the FS-75 fines were sieved through a 2.36 mm mesh size screen. Once sieved, the material was prepared further by mixing the components with water in a Hobart laboratory mixer. After a double batch of mixture was prepared, the charge was passed through the shearing plate of the laboratory extruder. Half of one batch was subsequently agglomerated in the laboratory extruder, while the other half was subjected to souring for 4 h. The extruder was equipped with a steel flat die with 4 holes of ~ 20 mm in diameter. Samples were collected immediately from the die and tested to determine moisture content and density.

The samples were subjected to drop tests. The samples with a weight of about 500 g were placed in a plastic bag and dropped 12 times onto a concrete surface from a height of 1 m. The amount of the fines formed during the tests was measured. The fines consist of particles that are smaller than 4.75 mm in size.

The effects of forced drying versus prolonged drying under natural conditions were studied by placing one sample from each batch in a dryer at a temperature of 150 °C until the moisture content of the brex reduced to 1%. The rest of the batch was stored in a laboratory at constant temperature and humidity, which simulates natural storing conditions. The temperature in the laboratory is about 20 C throughout the year. When dried, the brex were subjected to drop testing and axial compression testing. The briquettes were tested after 24 h, 3 days and 8 days of storing. The properties of the brex and the test results are given in Table 4.18.

As indicated by the extrusion parameters, there are several significant differences between lignin bonded samples and samples with a starch binder. Electric current in the extruder reached its maximum when producing samples with lignin. As for samples with starch, electric current was much lower. This decline in the rate of electric current is mostly likely due to the difference in moisture content. The relative moisture content in extruded samples with starch is higher than that in the samples with lignin, which can be explained by the fact that starch has a higher water absorption capacity than lignin. Starch presumably serves as an internal lubricant in the mixtures, as evidenced by lower temperature of the extruded masses.

Table 4.18 Results of brex strength tests after 24 h, 72 h and a week

Properties of briquettes after 24 h-storing at 20 °C						
Moisture content (%)	3.4	3.4	3.5	3.2	6.3	4.6
Amount of fines after drop test (less than 4.75 mm in size) (%)	8.0	21.0	17.0	13.0	3.0	1.0
Compressive strength: (MPa)	0.26	0.23	0.40	0.31	0.83	0.63
Properties of briquettes after a 3-day storing at 20 °C						
Moisture content (%)	1.2	2.0	0.9	1.7	3.7	1.6
Amount of fines after drop test (less than 4.75 mm in size) (%)	2.0	11.0	1.0	3.0	1.0	2.0
Compressive strength: (MPa)	0.66	0.56	3.30	2.00	1.90	2.10
Amount of fines (less than 4.75 mm in size) after drop test of forced dried samples (%)	3.0	4.0	3.0	4.0	3.0	3.0
Compressive strength: (MPa) of forced dried samples	12.6	9.0	9.6	8.4	14.0	12.8
Properties of briquettes after a week- long storing at 20 °C						
Moisture content (%)	0.7	0.7	0.5	0.7	0.6	0.7
Amount of fines after drop test (less than 4.75 mm in size) (%)	0.5	1.2	0.7	1.3	0.6	0.7
Compressive strength: (MPa)	13.8	11.1	15.2	10.5	20.9	19.2

The compressive strength in all the samples that were forced dried was much higher than in those samples that underwent 3-day storage in natural conditions. A sharp increase in strength was observed after a week of storing. The strength of the samples after storing exceeded the strength of the samples that were forced dried.

The drop test results of freshly agglomerated samples show significant differences in strength between samples containing lignin and samples containing starch. The drop test results of the lignin-based samples vary in the 20–30% range, while these values were 1–2% for the samples with starch.

The drop test results of forced dried samples are acceptable for all samples. After 3 days of storage, samples containing lignin showed good results in the drop tests, except for sample No. 2 (11%). These results can be due to the samples relative moisture content at this stage of the test. Samples without preliminary souring had lower moisture content and greater impact strength, which indicates a low efficiency of this procedure when agglomerating crushed ferroalloys fines.

Samples with a starch binder showed consistent results in drop tests. Such brex can be of sufficient strength for transport. Moreover, the starch binder content in the brex is lower than in the above-mentioned Silingen Polska briquettes (3%), while starch was only a part of the binder in these briquettes (in addition to 2% of liquid glass, by weight), and the compressive strength was 14 MPa (42% lower than in brex).

Fig. 4.25 Brex production at
Aktobe Ferroalloys Plant
(Kazakhstan Republic)

The results of the above-mentioned studies led to the construction of several
SVE agglomeration plants in different countries of the world for the agglomeration
of manganese ore fines and baghouse dusts, chromium ore fines and aspiration
dusts as well as different anthropogenic materials generated during production of
ferroalloys. The first smelts already confirmed the justification behind our recommendation for the brex as an effective charge component for SEAF.

In 2017 SVE plant for production of brex made of chromium ore fines and
aspiration dusts (Fig. 4.25) was commissioned in Kazakhstan (TNC Kazchrome,
Aktobe). Brex are being successfully used as a charge component of SEAF for
ferrochromium smelting.

The decision to build this line was made according to the results of the laboratory tests. Raw materials for testing (chromium ore fines and aspiration dusts,
Fig. 4.26) and the bentonite were combined with water and mixed in a Hobart
laboratory mixer, then processed through the laboratory extruder under vacuum.
The intent was to simulate the typical extrusion scenario of processing material
through a pug sealer and then extruder. The brex composition was as follows:
chromium ore fines and aspiration dusts—97%; bentonite—3%. Figure 4.27 shows
the appearance of the green brex.

Fig. 4.26 Chromium ore fines mixed with aspiration dust of ferrochromium production

Fig. 4.27 Green brex made of chromium ore fines and aspiration dusts

Within one hour of extrusion, 500 g of each sample was placed in a sealed plastic bag and dropped four times from a height of 1 m. The results were then screened and the fraction of particles below 4.75 mm was determined (Fig. 4.28). Less than 1% of fines were generated during this testing.

Currently, the brex are used in the charge of SEAF of the company.

Fig. 4.28 Results of the drop test of brex made of chromium ore fines and aspiration dusts

4.5 Conclusions

1. Manganese ore concentrate and the baghouse dust can be efficiently agglomerated by the SVE agglomeration technology.
2. Brex has sufficient mechanical strength to withstand the distant transportation including multiple loading-discharging operations. The mechanical strength of brex is significantly higher even with the addition of relatively small amounts of the Portland cement binder than the strength requirements to the charge components of the Submerged EAF. Furnace baghouse dust is also exhibiting binder properties. A 30% or more increase in baghouse dust content reduces the mechanical strength of brex.
3. Brex have sufficient porosity to ensure good reducibility. Porosity level is comparable with that of commercial manganese ores.
4. Brex exhibit high thermal stability and sufficient hot strength ensuring the stable and smooth operation of the furnace.
5. Brex with the baghouse dust content up to 30% will exhibit sufficient specific electric resistivity comparable with the specific electric conductivity the lumpy ores.
6. Brex can be efficiently used as one of the essential charge component (up to 30% of the ore part of the charge) for the Silicomanganese smelting thus improving technical and economical parameters of the furnace and decreasing the self-cost of Silicomanganese production.
7. The hot strength of brex in the up to 700–800 °C temperature range is achieved by a cement binder, and the hot strength in 800–1250 °C temperature range is achieved by the density of the formed silicate phase.
8. SVE technology can be used in the production of competitive commercial brex made of the ferroalloys.
9. Ferroalloys fines and ferroalloys crushing unit aspiration dust can be efficiently agglomerated by means of the SVE method to produce a charge component of SEAF with acceptable metallurgical properties.

References

1. Tolochko, A.I.: Utilization of dust and residues in ferrous metallurgy. In: Tolochko, A.I., Slavin, V.I., Suprun, Y.M., Khairudinov, R.M. (eds.) 143p. Metallurgy, Moscow (1990)
2. Gasik, M.I., Lyakishev, N.P., Emlin, B.I.: Theory and Technology of Ferroalloys Production, 784p. Metallurgy, Moscow (1988)
3. Zhuchkov V.I., Smirnov L.A., Zayko V.P.: Technology of Manganese Ferroalloys, 442p. UrB RAS, Yekaterinburg (2008)
4. Duarte, A., Lindquist, W.E.: Recovery of nickel laterite fines by extrusion. In: Proceedings of 27th Biennial Conference, pp. 205–217. IBA, USA (1999)
5. Redl, C., Matthias Pfennig, R.K., Fritsch, S., Muller, H.: Refining of ferronickel. In: The Thirteenth International Ferroalloys Congress Efficient Technologies in Ferroalloy Industry, 229, Almaty, Kazakhstan, 9–13 June 2013

6. Igrevskaya L.V.: Trends in Development of Nickel Industry. World and Russia. Scientific World, Moscow, 268p (2009)
7. Zhdanov A.V.: Study of reducibility of manganese ore raw materials . In: Zhdanov, A.V., Zayakin, O.V., Zhuchkov, V.I. (eds.) Electrometallurgy, no. 4, pp. 32–35 (2007)
8. Sokolov, V.N., Yurkovets, D.I., Razgulina, O.V.: Determination of tortuosity coefficient of pore channels by computer analysis of SEM images. In: Proceedings of Russian Academy of Sciences, Physical Series, vol. 61, No. 10, pp 1898–1902 (1997)
9. Ivanova, V.P., Kasatov, B.K., Krasavina, T.N.: Thermal Analysis of Minerals and Rocks, 399p. Resources Publishing House, Leningrad (1974)
10. Tolstunov, V.L., Petrov, A.V.: Proceedings of Higher Institutions. Ferrous metallurgy. No. 4, pp. 9–14 (1989)
11. Glasser, F.P.: The ternary system CaO-MnO-SiO$_2$. J. Am. Ceram. Soc. **45**(5), 242 (1962)
12. Gasik, M.I.: Manganese, 608p. Moscow (1992)
13. Zhdanov, A.V.: Study of electric resistance of materials and batches used for ferromanganese production. In: Zhdanov, A.V., Zayakin, O.V., Zhuchkov, V.I. (eds.) Electrometallurgy, no. 6, p. 24 (2007)
14. Noskov, V.A., Bolshakov, V.I., Maimur, B.N. et al.: Pilot Production of Briquettes from Ferroalloy Screenings at OAO NZF (Nikopol Ferroalloy Plant, OJSC). Metallurgical and Mining Industry, no. 3, pp. 124–126 (2004)
15. Davey, K.P.: The development of an agglomerate through the use of FeMn waste. In: Proceedings of Tenth International Ferroalloys Congress, INFACON-X: Trans- Formation through Technology, Cape Town, South Africa, pp. 272–280 (2004)
16. Hycnar, J.J., Borowski, G., Józefiak, T.: Conditions for the preparation of stable ferrosilicon dust briquettes. Inżynieria Mineralna. J. Pol. Mineral Eng. Soc. **33**(1), 155–162 (2014)
17. Electronic resource http://www.silingen.eu/products/

Chapter 5
A Laboratory Study and Full-Scale Testing of Brex in Direct Reduction Iron (DRI) Production

5.1 Analysis of the Results of the Application of Brex in the Charge of a Midrex Reactor

The process of mini steel mills using oxidized pellets as a feedstock includes production of metallized pellets, which are used further for the production of direct reduction iron (DRI), production of steel in electric arc furnaces (EAF) and its continuous casting and rolling. Companies that are importing large amounts of iron ore pellets annually for DRI production and also during the course of the transportation, stockpiling, and charging of the pellets into metallization reactors, as well as the discharging of the metallized pellets, generate tens of thousands of tons of fine materials containing iron every year. Pellets fines and DRI sludge are usually dumped in piles and EAF dust and mill scale are sold to a third party. The results of a preliminary analysis indicate that the recycling of such materials in the form of briquettes would help to produce additional quantities of steel, and would also free up a significant area occupied by dumped wastes. Recovery of these wastes would generate additional revenues, surpassing revenue from direct sales of wastes.

We have studied the possibility of using stiff vacuum extrusion to produce agglomerated products suitable for use as charge components in DRI reactors. The experiments were conducted in an industrial Midrex reactor operated by one of the customers of the J.C. Steele&Sons, Inc. This company imports 3.5 million tons of iron ore pellets annually to produce 2.35 million tons of hot-briquetted iron (HBI). Tens of thousands of tons of finely dispersed iron-bearing wastes are formed each year during the unloading, storage, and subsequent charging of pellets into metallization reactors, and also during the discharging of the metallized pellets, and their briquetting.

© Springer International Publishing AG 2018
I. Kurunov and A. Bizhanov, *Stiff Extrusion Briquetting in Metallurgy*,
Topics in Mining, Metallurgy and Materials Engineering,
https://doi.org/10.1007/978-3-319-72712-7_5

5.1.1   Tests with Brex in a Rigid Steel Basket in the Charge of a Midrex Reactor

We conducted two series of basket tests that entailed the reduction of brex in an industrial Midrex reactor. In the first series, 25 to 30 strengthened brex were placed inside a rigid steel basket (Fig. 5.1a) that were then charged into the reactor together with pellets. The brex were extricated from the reactor after the tests were concluded in order to be able to visually evaluate the condition of the reduced briquettes, and study their composition and properties. The mechanical strength of the brex did not play a significant role in this case, since they were not subjected to pressure from the column of charge materials.

The stiffness of the basket completely prevented the brex from being subjected to pressure from the layer of pellets and thus possibly be deformed or fractured. In the second series of tests, brex were placed inside deformable gas-permeable steel packets (Fig. 5.1b), which made it possible to study their behavior under conditions similar to those encountered inside a layer of pellets in a Midrex reactor. The results of the first series of tests are discussed below.

To conduct the tests, we used brex produced from a mixture of pellets fines— 55.6% (no more than 92% of which were 6.3 mm or smaller in size), metallized DRI sludge—27.8% (also with a 92% particle content no coarser than 6.3 mm), and mill scale—16.6% (with 99% no coarser than 10 mm). The main goals of the investigation were to evaluate the reduction properties of brex in the Midrex process and to select a binder that would give the brex the necessary strength and maximize their metallization. The mixture was subjected to additional pulverization (Fig. 5.2) on a roll crusher before it was briquetted. Table 5.1 shows the chemical composition of the components of the mixture.

Four types of experimental brex were tested (Table 5.2). The types differed negligibly from one another in terms of their chemical composition and differed only in the type, and the amount of binder that was used. The brex was made by J.C. Steele&Sons on a laboratory extruder. The mixture was prepared and

(a)   (b)

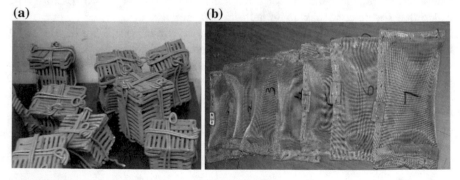

Fig. 5.1 Steel baskets (**a**) and packets (**b**) for charging brex together with pellets into a Midrex reactor

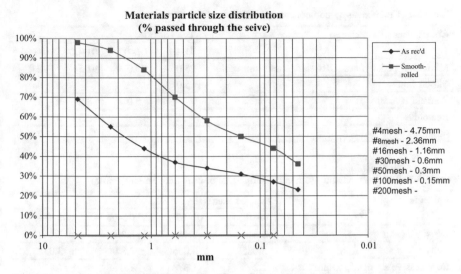

Fig. 5.2 Granulometric composition of the mixture of materials for briquetting: (1) initial mixture; (2) after additional pulverization (the maxima on the curves correspond to the mixture's content of particles smaller than 5 mm)

Table 5.1 Chemical composition of the components of the mixture for brex production

Elements and oxides	Content in the component, wt%		
	Pellets fines	DRI sludge	Mill scale
Fe	65.0	66.6	70.0
$SiO_2 + CaO$	3.8	6.52	1.00
CaO	1.3	4.38	0.15
MgO	0.75	0.69	0.10
Al_2O_3	0.95	0.83	0.25
MnO	0.1	0.16	1.20
P	0.055	0.045	0.02
TiO_2	–	0.010	0.020
V_2O_5	–	0.12	0.025
C	–	–	0.30
S	0.015	0.01	0.015
$Na_2O + K_2O$		0.33	–

homogenized by using a Hobart laboratory mixer that simulated the processing of the charge in a pug sealer.

There was a substantial difference (Table 5.3) between the green brex and the brex after strengthening (that is to say after they had been strengthened over the course of 14 days).

Table 5.2 Component-by-component composition of the brex

Components of the charge	Type of brex			
	01–01	01–02	01–03	01–04
Mixture of pellets fines, DRI sludge and mill scale	95.0	95.0	91.5	92.0
Slaked lime	5.0			6.0
Portland cement		5.0	8.0	
Bentonite			0.5	
Molasses				2.0

Table 5.3 The physical and mechanical properties of brex after strengthening

Properties	Type of brex			
	01–01	01–02	01–03	01–04
Density[a], g/cm^3	3.314	3.458	3.300	3.464
Density[b], g/cm^3	3.085	3.005	2.844	3.038
Compressive strength[a], N/mm^2	2.2	1.6	1.6	18.4
Moisture content[a], %	11.5	10.5	9.8	8.9
Moisture content[b], %	1.11	1.07	1.06	2.53

[a]Green brex
[b]Strengthened brex

The porosity of brex is one of their most important properties and determines their reducibility. The microstructure of the specimens and the specifics of their porosity were investigated with the use of a scanning electron microscope (SEM) LEO 1450 VP (Carl Zeiss, Germany) with a guaranteed resolution of 3.5 nm, and STIMAN software [1] to quantitatively analyze SEM images obtained in the backscattered electron regime. The porosity of briquette specimens 01–01 and 01–03 was studied on material that was freshly cleaved from the specimens' surface. Morphological studies of the microstructure were performed in the secondary electron regime and allowed us to obtain high-quality halftone images over a broad range of magnifications. The STIMAN method makes it possible to obtain correct images with distinct boundaries between the pores and the particles. The morphological parameters of brex 01–01 and 01–03 that were measured in this way, together with the characteristics of the pores are shown in Table 5.4. This information clearly shows that the only significant difference in the porosity of the specimens was with respect to the anisotropy coefficient. This difference was probably due to the type of binder.

The distribution of the macropores (with sizes greater than 100 µm) can be determined by computer-assisted X-ray tomography. The images in Fig. 5.3 illustrate the difference in the distributions of the macropores in green brex specimens 01–01 and 01–03. The images were captured with a Yamato TDM-1000 X-ray computer microtomograph (Japan). The magnification was 32 and the resolution

Table 5.4 The morphological parameters of the pore microstructure in brex specimens

Type of brex	Parameters	Class of pore, μm					D_{max}	n_{im}, %	K_a, %
		D_1	D_2	D_3	D_4	D_5			
		<0.1	0.1–1.0	1.0–10	10–100	>100			
	N, %	0.7	13.5	56.2	29.6	–	32.36	21.25	8.50
	K_f			0.42–0.50; 0.58–0.67					
	N, %	1.9	18.2	46.9	33.1	–	32.2	20.88	17.69
	K_f			0.33–0.42; 0.50–0.58					

Notes (1) N represents pores of different size classes expressed as fractions of the total number of pores n_{im} calculated from the SEM image; D_1–D_5 are the different pore-size classes; D_{max} is the maximum equivalent diameter of the pores; K_a is the anisotropy coefficient, or the degree of orientation of the solid structural elements; K_f is the shape factor of the pores. (2) The coefficient K_f is calculated as the ratio of the semiminor of an ellipse inscribed in a pore to its semimajor axis. The coefficient $K_f = 0.66$–1.00 for isometric pores, $K_f = 0.1$–0.66 for anisometric pores, and $K_f < 0.1$ for slit-shaped pores

was 11 μm. It is possible to conclude that the 01–01 brex had a higher percentage of pores larger than 100 μm than the 01–03 brex.

The degree of open porosity determined by the STIMAN method is very consistent with the open porosity measured by liquid saturation in a vacuum in accordance with the standard DIN 51056 (GOST 26450.1–85). The porosity values were within the range 21–24% for the specimens that were examined.

Cracks could clearly be seen on brex specimens 01–02 and 01–03 after they were removed from the reactor (Fig. 5.4).

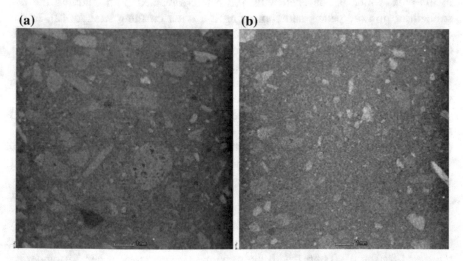

(a)   **(b)**

Fig. 5.3 Distribution of pores coarser than 100 μm in specimens of briquettes 01–01 (**a**) and 01–03 (**b**) (micro-X-ray tomography)

Fig. 5.4 The look of the reduced briquettes after their extraction from the rigid steel baskets

Both specimens had been prepared with the use of a cement binder. There were more cracks in the specimens with the 01–03 designation, which had higher binder content (Table 5.2). It is possible to conclude that the hot strength of brex is related to their crushing strength. Brex specimens 01–02 and 01–03, which were weaker and less dense in the cold state (see Table 5.3), also had a lower hot strength. The lower value for their hot strength manifested itself in the formation of surface cracks in the brex, which is consistent with the results obtained from studying the hot strength of iron ore pellets and comparing it to their crushing strength [2]. In the case of the pellets, there was also a negative dependence of hot strength on the ratio Al_2O_3/SiO_2. An increase in this ratio decreased the pellets' hot strength. Brex specimens 01–02 and 01–03, which exhibited a lower hot strength than brex prepared with a lime binder, had the highest value for elastic modulus due to the substantial amount of alumina in the cement and, in particular, in the bentonite (brex 01–03). At the same time, the presence of lime and hematite (pellet fines and scale) in brex 01–01 and 01–04 helped calcium ferrites form at just 400–500 °C during the course of the briquettes' heating in the Midrex reactor. These ferrites, which have low melting and softening points, strengthen the structure of brex and improve their reducibility [3].

After the brex were extracted from the rigid steel baskets, we determined their chemical composition, total iron content, content of metallic iron, and degree of metallization (Table 5.5). The highest degree of metallization was seen for the 01–01 brex with 5% lime content, while the lowest degree of metallization was registered for the 01–03 brex with the maximal (8%) content of cement binder.

Table 5.5 Chemical composition of the green and reduced brex

Type of brex	Green brex		Reduced brex			
	Fe_t, %	C, %	Fe_t, %	Fe_{met}, %	Metallization, %	C, %
01–01	64.53	2.32	80.07	77.39	96.65	0.98
01–02	64.06	1.88	79.22	74.63	94.21	0.84
01–03	62.53	1.72	75.20	68.37	90.92	0.87
01–04	60.46	2.60	76.40	70.17	91.85	1.70

Table 5.6 Volume and number of pores in brex specimens

Characteristics	Type of brex (green/reduced)	
	01–01	01–03
Total volume of the pore space, mm^3	0.77/0.45	0.69/0.87
Number of pores in the investigated volume of the specimen	186063/ 93321	198420/ 241634

The changes in the porosity of the brex during their reduction were evaluated by using a Phoenix V|tome|XS 240 X-ray microtomograph with two X-ray tubes: a micro-focus tube with a maximum accelerating voltage of 240 kV and a power of 320 W; a nanofocus tube with a maximum accelerating voltage of 180 kV and a power of 15 W. The initial analysis of the data and the construction of a three-dimensional model of the specimens based on the X-ray photographs (projections) were done using the datos|x reconstruction software, while the VGStudioMAX 2.1 and AvizoFire 7.1 software was used to visualize and analyze the data based on elements of the three-dimensional image. The photographs were taken at an accelerating voltage of 100 kV and a current of 200 mA; the resolution during the recordings was 5.5 μm. Table 5.6 shows the results obtained from measuring the volume and number of pores in the brex before and after reduction.

Figure 5.5 depicts the visualization of the pore distribution in the form of a 3D-model.

It is apparent that the porosity of the brex which were examined (and had coarse and ultra-coarse pores with a size of 1–100 μm or larger) changed in different directions. The total volume of the pore space decreased by a factor of 1.6 in the 01–01 brex and increased by 20% in the 01–03 brex. The increase in the porosity of the reduced brex with a cement binder can be attributed to the disintegration of the cement stone at 850 °C, and the accompanying decrease in the strength of the brex. The brex remained intact thanks to the formation of a metallic matrix composed of reduced iron. Conversely, the reduction of the brex with a lime binder was accompanied by a decrease in their porosity. This can be attributed to the agglutination of fine pores during the formation of the metallic matrix composed of reduced iron. Basket tests of brex made from a mixture of pellet fines, mill scale, and finely dispersed wastes from HBI production in an industrial Midrex reactor showed that they can be effectively metalized without loss of integrity. The brex

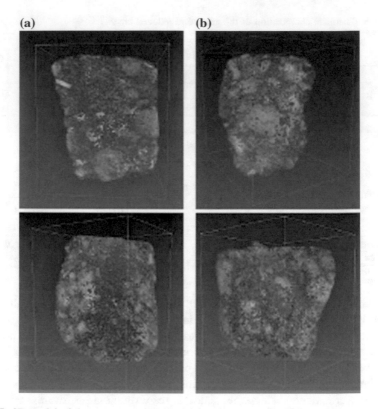

Fig. 5.5 3D-model of the pore space in a 01–01 brex (**a**) and a 01–03 brex (**b**)

with a lime binder displayed the highest degree of metallization and the greatest strength. While the brex with a cement binder remained intact, cracks were formed on their surface. The number of cracks was found to be greatest in the brex with a higher binder content.

5.1.2   Tests with Brex in Deformable Steel Packets in Charge of Midrex Reactor

At the second stage of testing, experimental brex were placed inside deformable steel packages that were gas permeable for the reducing gas (Fig. 5.1b). This input method allows brex behavior to be simulated adequately in actual reactor conditions, including the mechanical pressure of the surrounding traditional charge. At the end of the process, these packages were drawn out of the reactor, which enabled the condition of the reduced brex to be determined visually and their reduced chemical composition and properties to be explored.

Table 5.7 Chemical composition of the mix components

Chemical compounds	Pellets fine	Mill scale	DRI sludge	EAF dust
Fe_{tot}	65.00	70.0	66.2	29.68
SiO_2	2.50	1.00	2.14	4.25
CaO	1.30	0.15	4.38	19.74
MgO	0.75	0.10	0.69	24.27
Al_2O_3	0.95	0.25	0.83	1.32
MnO	0.10	1.20	0.16	0.96
S	0.015	0.015	0.01	0.13
$Na_2O + K_2$	0.034	–	0.33	1.42

The prepared mix of pellets fines (52.6%), metallized sludge from DRI production (26.4%), mill scale (15.8%), and EAF dust (5.2%) has been used.

The chemical composition of the mix components is given in Table 5.7.

All test samples were extruded using a 25 mm round pelletizing die. A Hobart laboratory mixer was used to simulate the mixing with water and pugging of the ground feed material in the open tub of the pug sealer. The laboratory extruder simulates the processing of the material through the sealing auger and die, into the vacuum chamber, and then final extrusion. The laboratory extruder consists of two chambers with a sealing die between them. The rear chamber is fitted with a 3-in. diameter sealing auger that pushes material through the sealing die. The second chamber can be subjected to a vacuum. It is also fitted with a 3-in. diameter auger that extrudes material through a pelletizing die. A PC-based data system monitors and records extrusion data. All mixes were extruded immediately after mixing. Three types of the experimental brex were produced with the compositions given in Table 5.8.

A magnesium sulfate-based binder was composed using the magnesium sulfate heptahydrate ($MgSO4·7H2O$). The binding properties of magnesium sulfate were first described by Zhuravlev and Zhitomirskaya [4]. The change in form that occurs when a dehydrated inorganic salt is converted to the hydrated crystal form is the basis for the recommendation that a partially dehydrated magnesium sulfate be used

Table 5.8 Experimental brex compositions, %

Charge component	Brex No. 1	Brex No. 2	Brex No. 3
Pellets fines	50.0	50.0	50.0
Sludge	25.0	25.0	25.0
Mill scale	15.0	15.0	15.0
EAF dust	5.0	4.75	5.0
Slaked lime	5.0	–	–
Portland cement	–	5.0	–
Magnesium binder	–	–	5.0
Bentonite	–	0.25	–

Table 5.9 Chemical composition of experimental brex before reduction, %

Elements and oxides	Brex No. 1	Brex No. 2	Brex No. 3
Fe_t	62.47	61.30	62.61
C	1.49	1.40	1.05
CaO	9.24	8.41	4.30
MgO	3.61	3.07	6.24
SiO_2	2.45	4.91	2.76
Al_2O_3	1.29	1.81	0.99
TiO_2	0.10	0.12	0.12
V_2O_5	0.076	0.07	0.08
MnO	0.34	0.36	0.35
P_2O_5	0.07	0.09	0.06
S	0.08	0.08	0.50
$Na_2O + K_2O$	0.19	0.93	0.83
Cl	0.04	0.03	0.02
ZnO	0.36	0.40	0.35

industrially as a binding material. Setting begins in 3 min and is complete in 6 min. Chemical composition of the brex is given in Table 5.9.

A calibrated electronic scale with a density measuring attachment was used to determine brex density. A compression tester was used for sample strength measurements, tensile splitting strength has been measured in accordance with ASTM (American Society for Testing and Materials) C1006-07, Paragraph 7.1. Porosity has been measured in accordance with DIN (Deutsches Institut für Normung) 51056. Moisture content was measured using a moisture balance. The physical and mechanical properties of the raw brex are given in Table 5.10.

Experimental brex with the above compositions were placed into the deformable and gas-permeable steel packages (each type of brex was placed in a separate package) and these packages were added to the traditional charge of the Midrex reactor. The tests were conducted during the course of a night shift from midnight to 8:00 a.m. The reducing gas temperature was around 900 °C. At the end of the process, the packages were removed from the reactor due to the presence of so-called cold discharge (with a discharge temperature of 40 °C) thus enabling the condition of the reduced brex to be visually determined (Fig. 5.6) and their chemical composition and properties to be examined.

Table 5.10 Physical and mechanical properties of raw brex

Property	Brex No. 1	Brex No. 2	Brex No. 3
Density, g/cm^3	3.5	3.48	3.66
Compressive strength, N/mm^2	4.8	11.1	4.4
Tensile splitting strength, N/mm^2	1.5	1.4	1.2
Porosity	29.7	24.7	25.1
Moisture content, %	8.4	8.4	8.6

(a) (b)

(c) (d)

Fig. 5.6 Raw brex (**a**) and reduced brex after their extraction from the steel packages **b** brex No. 1, **c** brex No. 2, **d** brex No. 3

An observation can be made that the brex that used a magnesium sulfate-based binder demonstrated the highest degree of integrity. The cement-bonded brex demonstrated the largest degree of fines generation. The brex that used lime binders were destroyed to a markedly lesser extent.

The mechanical strength values of the reduced and non-destroyed brex were defined in the course of splitting tensile testing (0.4 N/mm^2 for lime-bonded brex and 1.2 N/mm^2 for brex with a magnesium sulfate-based binder). The strength of the brex with a magnesium sulfate binder has not changed compared to its original value (Table 5.10). Thus, the brex with the lowest values for the cold mechanical strength showed the highest values for hot strength. Respectively, the brex with the magnesium sulfate-based binder also showed the highest reducibility. The decrease in the reducibility of the lime- and cement-bonded brex is related to the decrease in the permeability of the steel packages after the particles of the generated fines have blocked the holes. The chemical composition of the reduced brex and their degrees of metallization are presented in Table 5.11. One can see a serious discrepancy in terms of the total iron contents in the reduced brex. This, in turn, comes from a difference in metallic iron content. The reason for this is related to the different hot strength levels of the brex and to the aforementioned blocking of the holes by fines from partially disintegrated brex (No. 1 and No. 2), which prevents the intake of the reducing gas.

Table 5.11 Chemical composition and metallization of the reduced brex

Elements and oxides	Brex No. 1	Brex No. 2	Brex No. 3
Fe_t	74.86	69.02	86.86
Fe_{met}	49.11	18.66	84.00
Metallization, %	65.60	26.96	96.71
C	1.75	0.89	1.02
CaO	9.11	6.71	5.07
MgO	2.93	2.24	7.81
SiO_2	3.41	4.35	4.04
Al_2O_3	1.57	1.45	2.13
TiO_2	0.11	0.12	0.12
V_2O_5	0.07	0.08	0.07
MnO	0.34	0.43	0.39
P_2O_5	0.08	0.07	0.01
S	0.03	0.09	0.26
$Na_2O + K_2O$	0.48	0.51	0.74
Cl	0.09	0.01	–
ZnO	0.18	0.30	0.05

The Midrex reactor, which has been used for full-scale testing, produces DRI that has the following guaranteed parameters: a degree of metallization of 94%, total iron content—91%, metallic iron content—85%, carbon content—1.3%, sulfur—0.005%. It follows from the testing results that the brex with the magnesium sulfate-based binder reached a higher degree of metallization (96.71%), and almost the same level of metallic iron and carbon content, but this was smaller than in the DRI total iron content. The sulfur content is significantly higher in this brex. This is evidently a limiting factor for the utilization of brex with a magnesium sulfate-based binder in Midrex reactors. Sulfur will partially go to the flue gas and, in the absence of the flue gas desulfurization, it could spoil the quality of the catalyst. A lime-bonded brex has much lower sulfur content.

We can compare these results with those obtained in the Midrex reactor of the ArcelorMittal in Hamburg with roller briquettes of the following composition: 37.6% DRI filter cake, 47% oxide fines, 9.4% return fines, and 6% of hydrated lime and molasses as binders [5]. The total iron content in the raw briquettes amounted to 59%, in the reduced—77%. The degree of metallization of the briquettes extracted from deformable steel packages was 86% [6]. The brex with a magnesium sulfate-based binder revealed a greater degree of metallization (96.71%) with a lower initial value of the total iron content (62.61%) in the raw brex and with a smaller binder content (5%).

5.1.3 *Mineralogical Study of the Reduced Brex*

The structure of the reduced samples of brex has been studied using light microscopy methods with the "Nikon" equipment (a polarized ECLIPSE LV100 POL
microscope equipped with a digital photomicrography system DS-5 M-L1) (Nikon,
Tokyo, Japan). Analysis of mineral phases in polished sections of briquettes was
conducted by MLA 650 (FEI Company, Hillsboro, OR, USA), including an FEI
Quanta 650 SEM scanning electron microscope. The main mineral phases were
determined by a diffraction analysis using an analytical complex ARL 9900
Workstation IP3600 (combined X-ray fluorescence spectrometer) (Thermo Fisher
Scientific, Waltham, MA, USA). Reduced brex with a magnesium sulfate-based
binder and retained integrity samples of the reduced brex with lime and cement
binders were tested.

Fines from pellets with a low basicity ($CaO/SiO_2 = 0.3$–0.5) are represented by
fragments with a different mineral composition: cores in the form of residual
magnetite with a glass phase and shells consisting of splices of hematite with
calcium ferrites. The mill scale has a magnetite composition—a solid solution
$(Fe,Mn)O \cdot Fe_2O_3$. A special role in the composition of the brex is played by EAF
dust, which consists of the main melt generating components: CaO, MgO, MnO,
SiO_2, and Na_2O.

In the reduced and non-destroyed cement-bonded brex, iron is represented by
partly recovered iron-containing minerals made from pellets fines and mill scale.

In the major body of the cement-bonded brex extracted from the steel
mesh packages, there are the areas with the phases that are close in terms of their
composition to the binary metasilicates $2Na_2O \cdot CaO \cdot 3SiO_2$ (N_2CS_3) and
$Na_2O \cdot 2CaO \cdot 3SiO_2$ (NC_2S_3) without any presence of iron (Fig. 5.7).

N_2CS_3 melts incongruently [7] (melting starts at 1141 °C with the creation of
78.5% of the melt and 21.5% of NC_2S_3, and ends at 1203 °C). NC_2S_3 melts
congruently with an extensive field of primary crystallization. In the Midrex process
temperature range, only two eutectics exist (Fig. 5.8): 755 °C—N_3S_8 + NCS_5 +
S (N_2O—22%; CaO—3.8%; SiO_2—74.2%); 827 °C—N_2CS_3 + NC_2S_3 + NS_2
(N_2O—36.6%; CaO—1.8%; SiO_2—60.7%).

In the phases detected by the X-ray spectral microprobe analysis, CaO content
exceeds 10%, indicating that limited melt formation takes place in these areas. The
phase detected at temperatures ranging from 870 to 950 °C can be in an amorphous
state with low amounts of liquid phases, as illustrated in Fig. 5.9.

In the cement-bonded brex, the alkali silicate phase is predominantly associated
with the smallest, dusty fractions, which are contained in the alkali. The silicate
bonding is not fully developed in the sample's body. Optical study of the samples
of the reduced brex shows the pieces of dicalcium silicate (Ca_2SiO_4) crystals.

Lime-bonded brex have a higher degree of metallization and a deeper interaction
between charge components. Such samples are characterized by a variation in the
composition of iron–silicate melts in different parts of the body of the brex. The
most prevailing is the combination of the primary melt with iron. In this case,

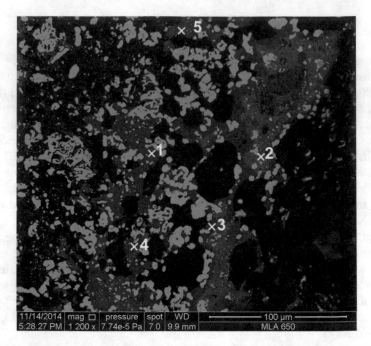

Fig. 5.7 SEM-image of the reduced cement-bonded brex. 1, 3, 4—binary metasilicates; 2—silica containing calcium ferrite; 5—dicalcium ferrite

Fig. 5.8 Diagram of the system Na_2O–CaO–SiO_2

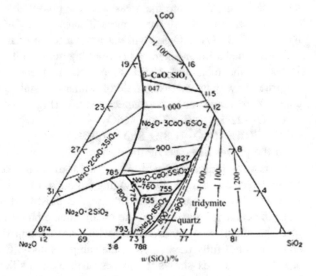

creation of the melt is related to the enveloping of the pellets fines particles by the dust fractions. The primary iron–silicate melt composition is close to olivine phases. The melt has a high basicity ($CaO/SiO_2 = 0.8$–0.9), Na_2O content in different parts

Fig. 5.9 Diagram of system
$Na_2O \cdot 2SiO_2–$
$Na_2O \cdot 2CaO \cdot 3SiO_2$

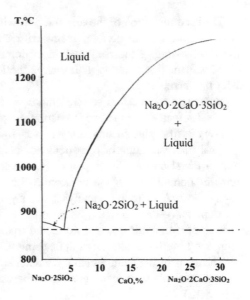

of the sample ranges from 1.0 to 10%, magnesium and manganese oxides content does not exceed 1.0–2.0 mass%. Under the conditions of the Midrex process as the pellets fines contact the melt, metallic whiskers are generated (Fig. 5.10).

Rarely are mineral formations with a high content of calcium oxide (up to 40%), silicon oxide content (25–26%), and the same amounts of bivalent iron observed in brex with lime as a binder. No other oxides are found in the composition of the amorphous iron–calcium phases with high calcium content. The composition of such phases is close to the Melilite phase ($Ca_2Fe_3 + Si_2O_7$).

Fig. 5.10 Generation of the metallic "whiskers" at the contacts of the pellets fines and melts

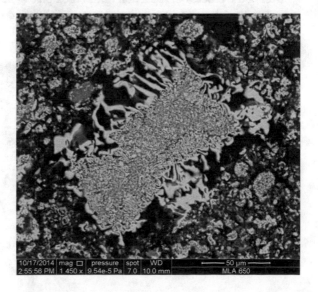

The predomination of the fine fraction within the brex is of particular interest. It consists of a combination of metal with calcium ferrites. Apparently, in pellets fines, consisting of the combination of hematite with calcium ferrite, the hematite phase was the first to reduce and such reduction took place without breaking contact with the ferrite.

Brex with magnesium sulfate binder exhibited the highest degree of metallization and a well-developed melt generation process.

For the first time in this study it was found that, during the metallization, the original microstructure of brex does not persist. Brex is close to a two-phase system: a metal and a silicate phase. However, the composition of the silicate part remains nonuniform in the adjacent volumes of the reacted fines. According to X-ray spectral microprobe analysis, among the studied melts, two mineral species are dominant—close to the olivine and Melilite structures, but each with a different content of iron oxides, magnesium, and manganese. This conclusion has also been confirmed by the results of an optical microscopy of the brex samples (Fig. 5.11).

The areas of the melt, which have reached equilibrium, are of particular interest. The mineral phase in combination with residual melt is observed in these areas (Fig. 5.12).

Thus, it is clear that the mechanism of the strengthening of brex samples in each case has its own specific characteristics. For cement-bonded brex, a lack of hot strength during heating in the reducing atmosphere is associated with their swelling and with the limited formation of silicate bonding. The strength of lime-bonded brex during the reduction process is provided by formation in the solid phase of the iron–calcium silicates, similar in composition to the olivine phase with a low melting temperature. Brex with the magnesium sulfate-based binder has the highest degree of generation of the two-phase metal–silicate system at the end of the metallization process.

During the process of reduction of the brex placed in deformable steel packages, their porosity changed in a different way. With the aim of studying the nature of

Fig. 5.11 Optical microscopy of the reduced lime-bonded brex (1—iron, 2—Mellilite; magnification 200)

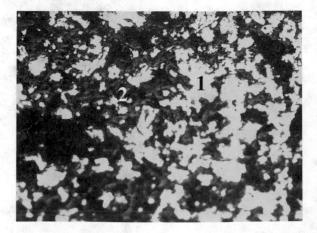

Fig. 5.12 Structure of the reduced brex (1, 2—mineral phase; 3, 4—residual melt)

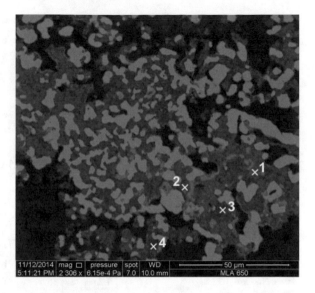

such changes, we have investigated the structure of the porosity of the original and reduced brex. The microstructure of samples was studied using a scanning electronic microscope (SEM) 1450 LEO VP (Carl Zeiss, Jena, Germany) with a resolution of 3.5 nm. For morphological studies of the microstructure of the samples, the secondary electrons mode has been used, which allows high-quality halftone images to be obtained across a wide range of magnifications. Quantitative analysis of microstructure was conducted using STIMAN software by two different methods: The first—is based on a series of SEM images obtained in backscattered electrons, which enables the correct picture to be obtained with clearer boundaries between the pores and particles; the second—by a complex analysis based on a set of SEM images, and images captured using Yamato TDM-1000 (Yamato Scientific, Tokyo, Japan) computed tomography. The results of the morphological studies are given in Table 5.12.

During the process of metallization, the total porosity of brex with a magnesium sulfate binder increased by 19.5%, cement-bonded brex by 13.15%, and brex with lime as a binder only by 2.6%. The porosity increased due to large and mega-pores (larger than 100 µm): for brex with magnesium sulfate binder—they increased by 6.9 times, and for lime-bonded brex—they increased by 2.43 times. In cement-bonded brex, the growth of share of large share was 1.24 times. The maximum diameter of pores in brex with a magnesium sulfate binder has increased by almost three times and in brex with a lime—by 1.23 times. The difference in the nature of changes of porosity between the original and reduced brex can also be clearly seen on images captured by computed tomography (Fig. 5.13). In brex with magnesium sulfate- and lime-based binders, the development of internal cracks took place. This contributed to the formation of coherent porous systems and the increase in the reduction rate.

Table 5.12 Morphological parameters of the original and reduced brex*

Brex	Porosity SEM/SEM + CT	Contribution of pores of different dimension categories to total porosity (ptot), %. Diameter in microns					Maximum diameter, μm
	p_{tot}, %	D_1	D_2	D_3	D_4	D_5	D_{max}
		<0.1	0.1–1.0	1.0–10	10–100	>100	
Brex No. 1	31.6	0.6	8.6	25.6	65.2	0.0	57.6
	37.3	0.5	7.2	21.4	66.8	4.1	407.2
Brex No. 1 (reduced)	32.9	0.2	9.0	38.4	52.4	0.0	70.94
	38.3	0.1	7.7	32.4	49.8	10.0	504.49
Brex No. 2	31.2	0.6	8.0	30.1	61.3	0.0	45.78
	35.7	0.5	7.0	26.2	61.3	5.0	299.96
Brex No. 2 (reduced)	37.9	0.2	9.3	31.1	59.1	0.3	100.51
	40.4	0.2	8.9	29.5	55.2	6.2	583.77
Brex No. 3	31.8	0.6	5.8	28.4	65.2	0.0	62.92
	32.9	0.5	5.5	27.2	65.3	1.5	191.89
Brex No. 3 (reduced)	32.8	0.2	3.2	35.7	60.9	0.0	71.75
	39.3	0.2	2.6	28.9	58.0	10.3	736.41

*p_{tot}—total porosity calculated by the analysis of SEM image; D_1–D_5—different dimension categories of pores; D_{max}—maximum pore diameter; CT—computed tomography

It can be assumed that the driving mechanism for the development of porosity in the brex was the crystal lattice type changes during the hematite–magnetite phase transformations, which results in an increase in the volume of the brex. This increase in volume is accompanied by mechanical stresses, which may lead to the partial destruction of brex, which leads to an increase in the proportion of large and mega-pores and to the formation of internal cracks. The stronger the brex is mechanically, the less it is prone to such partial destruction.

The brex with the magnesium binder were further manufactured at the industrial SVE briquetting line belonging to Suraj PL (Rourkela, India). Phase analysis of the industrial brex was done in Rikagu Miniflex600 X-ray diffraction system using copper target and nickel filter. Figures 5.14, 5.15, 5.16, 5.17, and 5.18 show the phase composition of the brex after reduction at different temperatures. The testing has been performed with the help of the experimental facility described in Chap. 4 (Figs. 4.2 and 4.3) and adjusted for the conditions simulating blast furnace process. Heating of the brex samples in a reducing atmosphere was carried out to temperatures of 1000, 1100, 1200, 1300, and 1400 °C.

Mineralogical study of the reduced samples was done on polished section under reflected light in Leitz Universal microscope with the image analyzer. Figures 5.19, 5.20, 5.21, 5.22, and 5.23 show the different stages of the brex metallization.

The results of this study show clearly that magnesium bonded brex could be considered as candidates for the BF charge.

(a)

(b)

(c)

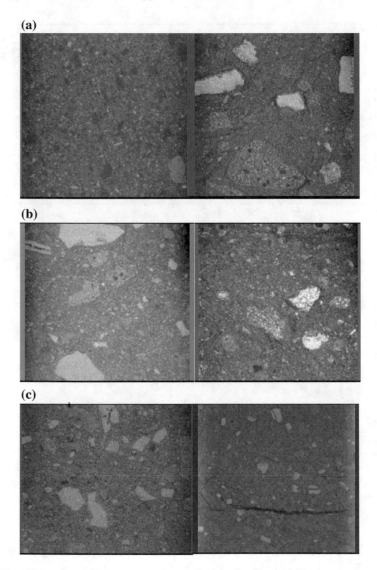

Fig. 5.13 Distribution of pores in the original (left) and reduced (right) brex; **a** lime-bonded; **b** cement-bonded brex; **c** brex with magnesium sulfate-based binder, magnification 64

As an appendix to the study of the properties of the brex as a charge component in Midrex reactors, we studied the properties of brex made from direct reduced iron (DRI) fines, that is formed in the course of the classification process designed to classify the pellets and the hot-briquetted iron that are manufactured at the Stock Company «Lebedinsky GOK» (Belgorod region, Russia). The entire metalized product is subjected to screening in terms of its classification: less than 4 mm and less than 25 mm. In order to obtain a competitive and economically viable product,

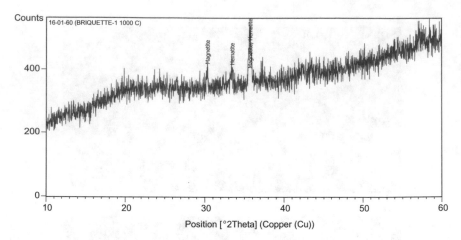

Fig. 5.14 Phase composition of brex sample after reduction at 1000 °C. Main phases: magnetite, hematite

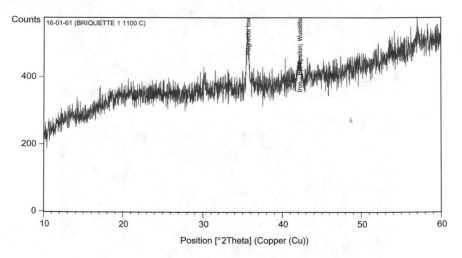

Fig. 5.15 Phase composition of brex sample after reduction at 1100 °C. Main phases: magnetite, wustite, iron

the brex must be used as a charge component in an arc furnace, and correspondingly its size, density, and binding agent should be such that the briquette does not float in the slag, but descend down to the melt and does not bring any components that are harmful to the steel. The durability should be sufficient to prevent any destruction during the course of transportation (either by vibration, or friction), and during its storage in the stockpile, or in the hold.

From the DRI fines formed at the «Lebedinsky GOK», brex with the following composition were prepared and tested by ourselves (Fig. 5.24): DRI fines 92%, molasses—4%, lime—4%, bentonite—0.5% (in excess of 100%).

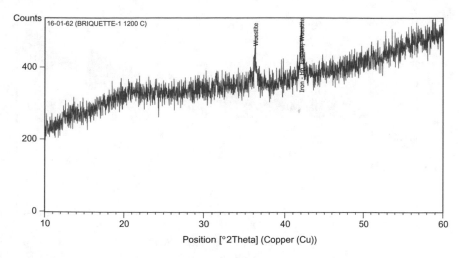

Fig. 5.16 Phase composition of brex sample after reduction at 1200 °C. Main phases: iron, wustite

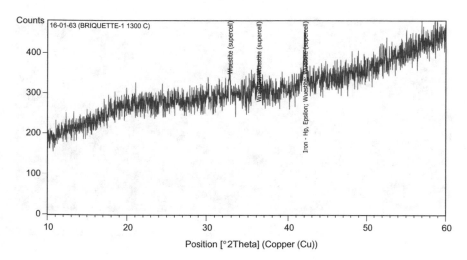

Fig. 5.17 Phase composition of brex sample after reduction at 1300 °C. Main phases: α-iron, wustite

The value for the crushing strength of the brex was on average 122 kg/cm². The porosity of the brex was 27.36%.

Thermal testing in the Tamman furnace during the melting of the brex at a temperature of 1513 °C demonstrated their high thermal stability. The brex melted smoothly and was immersed fully into the melt. The duration of the melt of a single brex was 10 min.

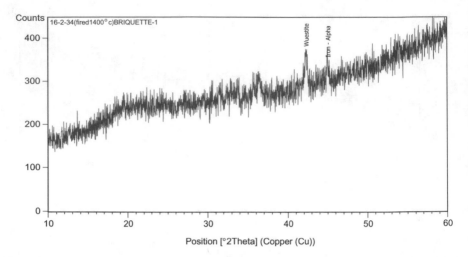

Fig. 5.18 Phase composition of brex sample after reduction at 1400 °C. Main phases: α-iron, wustite

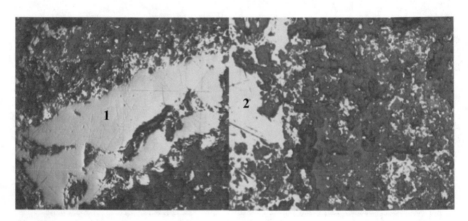

Fig. 5.19 Photomicrograph of Briquette-1 fired at 1000 °C 1—magnetite; 2—hematite; magnification ×200

5.2   High Temperature Reduction of Brex

We have investigated the behavior of the iron ore and coal brex in a neutral atmosphere at a temperature range of 1350–1370 °C partially simulating the conditions of the reduction of the iron ore and coal pellets in the ITmk3 process developed by «Kobe Steel» [8, 9]. This process is implemented in the annular chamber rotary hearth furnace (RHF). Green (unburned) iron ore and coal pellets are loaded in one or two layers onto a rotating hearth covered by a refractory. The heating of pellets and reduction of iron oxides is performed per single revolution of

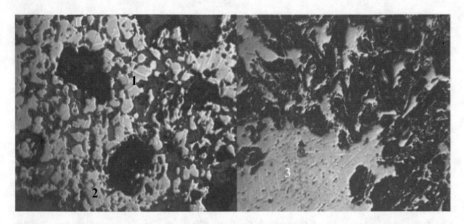

Fig. 5.20 Photomicrograph of Briquette-1 fired at 1100 °C 1—iron; 2—wustite; 3—magnetite; magnification ×200

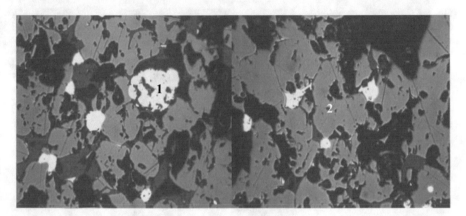

Fig. 5.21 Photomicrograph of Briquette-1 fired at 1200 °C 1—iron; 2—wustite; magnification ×200

the hearth over the course of 10–15 min at a maximum temperature that is higher than 1350 °C. Iron oxides are reduced by carbon in the iron ore and coal pellets or briquettes. The reduced iron in the pellets or briquettes is carbonized with its particles being coagulated and melted to produce bean-shaped cast iron drops (3–12 mm) inside the softened slag shell. When cooled in a neutral atmosphere these drops harden forming cast iron pellets (or "nuggets"). The metallurgical properties of pellets or briquettes must maintain their integrity during heating, and reduction and the slag shell that is produced should prevent the spread of liquid droplets of iron.

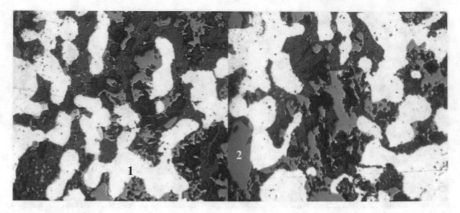

Fig. 5.22 Photomicrograph of Briquette-1 fired at 1300 °C 1—iron; 2—wustite; magnification ×200

Fig. 5.23 Photomicrograph of Briquette-1 fired at 1400 °C 1—iron; magnification ×200

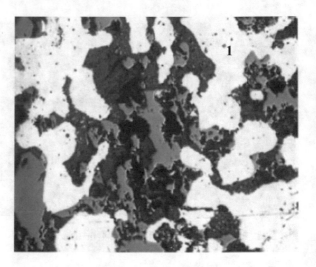

For the production of the experimental brex, we have used the hematite iron ore concentrate (63.13% of iron content; $Fe_2O_3 = 90.18\%$, $SiO_2 = 4.63\%$, Al_2O_3—3.11%, MnO—0.807%, CaO—0.346%), the Globe coal (bulk density 768 kg/m³; moisture content 7.8%; ash 4%; volatiles 37.7%; and where 73% of particles are less than 0.6 mm in size), and the Shubarkol coal (bulk density 800 kg/m³; moisture content 10%; ash 7.1%; volatiles 38.5%; S 0.40%; P 0.021%.; and where 99% of the particles are less than 0.3 mm in size). The Type I Portland cement (general use) has been used as the binder. Volclay DC-2 Western (Sodium) Bentonite was used as the binder and plasticizer. We have manufactured brex with

Fig. 5.24 Brex made from DRI fines (at the «Lebedinsky GOK»)

both round and oval cross sections (with diameters of 1/2″ and 1″; and a length equal to 1.5–2.0 diameters). The laboratory extruder simulates the processing of the material through the feeder to the pug mill and then to the sealing auger and die, into the vacuum chamber, and then final extrusion. Moisture content was measured using a moisture balance. A calibrated electronic scale with a density measuring attachment was used to determine brex density (Mettler MS603S and Mettler MS-DNY-43).

The results of the sieve analysis of Globe coal and Shubarkol coal are set out in Fig. 5.25.

Table 5.13 shows the chemical composition of iron ore concentrate (w-content).

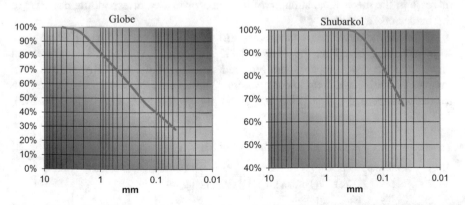

Fig. 5.25 Sieve analysis: left—Globe coal (73% particles less than 0.6 mm in size); right—Shubarkol coal (99% particles less than 0.3 mm in size)

Table 5.13 Chemical composition of iron ore concentrate

Component	w, %
Fe_2O_3	90.18
SiO_2	4.63
Al_2O_3	3.11
MnO	0.807
CaO	0.346
MgO	0.337
Na_2O	0.163
P_2O_5	0.129
TiO_2	0.091
S	0.050
BaO	0.026
CuO	0.022
ZnO	0.021
NiO	0.017
Cl	0.016
CO_3O_4	0.015

Table 5.14 gives the brex compositions, shapes, and their physical properties. Wet base moisture of the brex No. 1 and No. 2 was 18.5%. For the brex No. 3, moisture content was equal to 15.2%. The vacuum applied for the production of the brex No. 1 and No. 2 was at the level of 48.26 mm Hg and for the brex No. 3 the value was 38.10 mm Hg.

For the production of the brex No.3, we have applied a 24-h souring using Bentonite. This procedure enhances the plasticity of the mix and can lead to an increase in the mechanical strength of the brex (or to the corresponding decrease of the binder consumption).

We have measured the compressive strength of the cured brex. These values were measured based on the Instron 3345 device (USA). For the measurement procedure, we have prepared the required amount of the specially shaped brex samples and subjected them to compression. The values of the compressive strength for the brex were as follows (kgF/cm^2): sample No. 1—59.0; sample No. 2—24.0;

Table 5.14 Brex components, shape, and physical properties

Brex sample	Component, %				Density, g/cm^3	Porosity, %	CCS, kgF/cm^2
	Iron ore	Coal	Portland cement	Bentonite			
No. 1*	63	32	5	–	2.072	34.8	59.0
No. 2	63	32	5	–	2.034	36.9	24.0
No. 3	63.5	31.3**	4.7	0.5	2.207	36.4	49.0

*Oval shape (1.25 cm × 2.5 cm)
**Globe coal

sample No. 3—49.0. We have observed the influence of the shape of the brex on the mechanical strength. For the brex No. 1 and No. 2 with the same chemical composition, we attained a significant increase in the compressive strength for the oval-shaped brex when compared with the regular cylindrical form of the brex. As far as the values for the tensile splitting strength are concerned, this increase takes place when the applied pressure is perpendicular to the main axis of the oval. When the applied pressure is perpendicular to the short axis of the oval, we observe a corresponding decrease in the strength value. It is clear, however, that the most probable orientation for the oval-shaped brex lying on its side is with its long axis parallel to the flat surface. The mathematical and physical modeling of this phenomenon has been described and studied previously (Chap. 2). It is also worth mentioning that the transition to the oval shape increases the value of the external brex surface when compared with the round brex of the same length and volume.

We expected that the increased density of the brex might prevent their carbothermic reduction. However, the results of the porosity study show that for all brex compositions that were considered, the value of the porosity favors their reducibility. We used an LEO 1450 VP (Carl Zeiss, Germany) Scanning Electronic Microscope with a resolution of 3.5 nm together with the X-ray computed high-resolution computed tomography system Phoenix V|tome|X S 240 (General Electric, USA). Computed tomography has been used to detect the share of the macropores (larger than 100 μm in size). For the smaller size pores, the SEM has been applied. The approach based on STIMAN [1] computer software has been used to calculate the porosity detected by SEM. The results of these measurements showed that the total porosity of the brex No. 1 was equal to 34.8% (10.9% measured by X-ray computed tomography and 23.9% by SEM) and brex No. 2 was equal to 36.9% (12.3% measured by X-ray computed tomography and 24.6% by SEM). For the brex No. 3, the same value was equal to 36.4% (10.8 and 25.6% correspondingly). The spatial distribution of the pores detected by the X-ray computed tomography is illustrated by the image given in Fig. 5.26.

The reduction tests of the iron ore and coal brex following the conditions of the ITmk3 process were conducted by ourselves in laboratory resistance electric furnaces with a controlled atmosphere based on the Electric Furnace SSHVE-1.2, 5/25-I2 (Russia) with a vertical arrangement of the graphite heater and with an inside diameter equal to 65 mm. We positioned a working removable Alumina Crucible (98% Al_2O_3) in the isothermal zone of the heater. We regulated and stabilized the temperature in the furnace with the help of the thermocouple BP (A) 5/20 located in the isothermal area of the furnace on the outside of the heater. We measured the true value of the temperature using the thermocouple 2 BP (A) 5/20, located inside the heater and lowered from above into the working Alumina Crucible. According to available data on the process of ITmk3, we have selected 1360 °C as the working temperature, which we maintained with an accuracy of ±10 °C (1350–1370 °C).

Before the test, the furnace with the working replaceable Alumina Crucible was pumped out using the forevacuum pump until the residual pressure reached a value of 10^{-1} Pa and was then filled with a high-clean grade argon. Then, we opened the gas discharge to atmosphere and kept the argon consumption through the furnace at

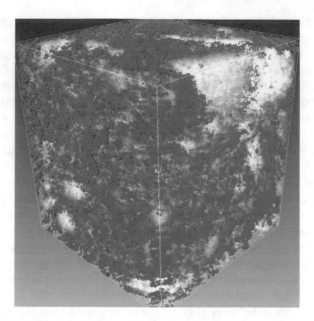

Fig. 5.26 Spatial distribution of the pores in brex No. 2 (X-ray computed tomography; dark and black—pore space; light and transparent—brex body; resolution 6 μm)

the level of 0.5 l/min. We switched on the heating of the furnace, reached the desired temperature level and stabilized it, and then dropped the brex sample through the dosing gateway into the crucible. This very moment was considered as the test start of the test. Thus, we have subjected the sample brex to the rapid heating from the ambient temperature to the working temperature of the furnace. During every single test, we recorded the process on a video using thermal imaging infrared camera «Pyrovision M9000» ("Micron", USA). This device records the film and also shows the temperature at each point of each frame. Sample exposure time was around 15 min; if necessary, we extended it up to 20 min. After that we depressurized the furnace, extracted the crucible with the products and tempered them in air, installed a new working crucible into the furnace and the new sample brex into the dosing gateway. We closed the furnace, washed it out using argon, and after setting the required temperature, we repeated the cycle.

Figures 5.27 and 5.28 show the photos of the consequent stages of the brex melting (volatiles release, solid-state reduction, melting and hot metal, and slag drop creation) based on the visual data obtained by «PyrovisionM9000».

In all cases, the high density of briquette and large amounts of volatile coal did not interfere with the solid-state reduction of iron. During the heating, the cracks at the brex surface appeared however the brex did not decompose and kept its integrity (Fig. 5.29). Finally, the brex was transformed into liquid metal (nuggets) and slag drops. In the event of an excess of reducing agents, we observed that the formation of metallized shells mixed with drops of slag.

Figure 5.30 shows the appearance of the products of the brex reduction.

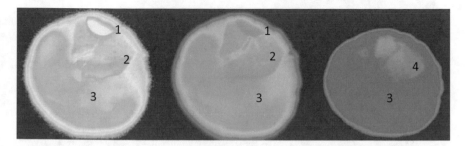

Fig. 5.27 Brex No.1 (from left to right: volatiles release, $t = 1$ min, $T = 1000$ °C; beginning of the melting, $t = 6$ min, $T = 1200$ °C; before complete melting, $t = 7$ min, $T = 1360$ °C); 1—thermocouple; 2—brex; 3—carbonaceous bed zone; 4—melting brex (solid and liquid)

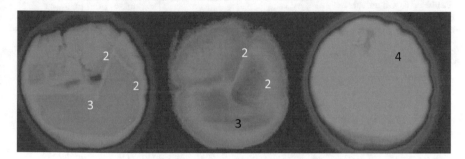

Fig. 5.28 Brex No. 2 (from left to right: the end of the coal volatiles release, $t = 1$ min 40 s, $T = 1000$ °C; solid-state reduction, $t = 6$ min 37 s, $T = 1200$ °C; creation of the nuggets, $t = 13$ min 20 s, $T = 1360$ °C; 2—brex; 3—carbonaceous bed zone; 4—liquid nugget surrounded by slag

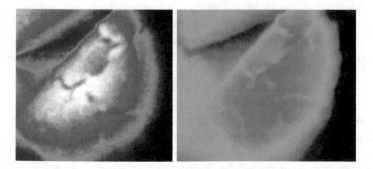

Fig. 5.29 Brex No. 3 at temperatures $T = 1100$ °C, left; and $T = 1250$ °C, right

Chemical composition of the reduced products (nuggets and shells) is given in Table 5.15. Due to the high content of coal in the brex, the carbon concentration in the metal was at the level of 3.3–4.8%. Metal samples (nuggets and shells) were

Fig. 5.30 Products of the brex reduction. From left to right: brex No. 1, brex No. 2, brex No. 3

Table 5.15 Chemical composition of metal (nuggets and shell)

Element, %	Sample No. 1	Sample No. 2	Sample No. 3
[Fe]	95.53	96.05	94.08
[C]	3.740	3.333	4.800
[Si]	0.252	0.154	0.605
[Mn]	0.076	0.121	0.136
[S]	0.107	0.118	0.185
[P]	0.135	0.058	0.085
[Cu]	0.0030	0.0074	0.0090
[Co]	0.0905	0.1030	0.0343
[Ti]	0.0580	0.0656	0.0570
[Zn]	0.0060	0.0005	0.0050

studied using a Jeol JSM 6490 LV Scanning Electron Microscope. All samples except the samples No. 3 have a similar microstructure: small island-type non-metallic inclusions in a metal matrix. The smallest size of inclusions is observed in the sample No. 2 (10–20 μm), in the sample No. 1, the inclusions are 50–60 μm in size. There are no oxide inclusions in the No. 2 sample. Sample No. 3 is characterized by nonhomogeneous inclusions of a linear form with a width of around 5–10 μm and a length of 200–300 μm. As can be seen from Table 5.15, the C and Si content are high in sample No. 2. This shows that the reduction material in sample No. 2 is excessive compared to samples No. 1 and No. 3. The decrease of the amount of the reduction material in sample No.2 could help to improve the strength of the brex. Synchronized thermal analysis methods and Mössbauer spectroscopy have been applied for the study of the phase composition of the brex (STA 449 C, Germany). A comparison of the peaks of the metallic iron formation shows that the reduction of the metal iron takes place to a larger extent in the brex samples No. 1 and No. 2 (with the Shubarkol coal) rather than in the brex No. 3 which contains Globe coal. The degree of fineness of the Shubarkol coal particles might have influenced their enhanced reducibility. For a definitive conclusion on

Fig. 5.31 Images of metal samples in backscattered electrons (from the left to the right—No. 1; No. 2; No. 3)

the suitability of the brex, it is necessary to conduct further research on optimization of their size and on the composition of the fluxing oxides in the brex.

Metal samples (nuggets and shells) were studied using a Jeol JSM 6490 LV Scanning Electron Microscope. Figure 5.31 shows the images of the samples in the backscattered electrons.

5.3 Conclusions

1. The research into the properties of brex made from dispersal wastes from mini-plants, producing rolled products, and using oxide pellets as a raw material and the study into the behavior of this brex as they are heated in a reducing atmosphere in an industrial Midrex process reactor enable a conclusion to be made—that SVE technology makes full and effective recycling of these above wastes possible.
2. The difference in the nature of the changes in the porosity of both original, and reduced brex depending on the type of binder used, has been established. In brex that used a magnesium binder and which retained their tensile strength, and did not form any fines, the metallization process is accompanied by the formation of a metallic carcass, and the small pores disappear, while the larger pores increase in size.
3. Brex that use a magnesium binder have a high degree of metallization after reduction, in the course of this process, the microstructure in the original brex is not retained and begins to resemble the condition under the two-phase «metal–silicate phase» system.
4. One limitation on the use of brex that employ a magnesium binder is the increased sulfur content in the binder, which is partially transformed into a flue gas, and in the absence of desulfurization of the flue gas, this reduces the quality of the catalyst. It is this same parameter that will also define the permitted content of these brex in the charge of a metallization furnace.
5. SVE can be used effectively to agglomerate HBI fines.

6. Brex made from coal and ore could be considered as a possible alternative to the coal and ore pellets for use in the ITmk3® process.
7. By way of a final conclusion on the suitability of these brex, research needs to be undertaken into the optimization of the dimensions and composition of slag forming oxides in the brex.

References

1. Sokolov, V.N., Yurkovets, D.I., Razgulina, O.V., Melnik, V.N.: Study of characteristics of solids microstructure by computer analysis of SEM images. In: Proceedings of Russian Academy of Sciences. Physical Series, vol. 68(9), pp. 1332–1337 (2004)
2. Dwarapud, S., Ranjan, M.: Influence of oxide silicate melt phases on the RDI of iron ore pellets suitable for shaft furnace of direct reduction process. ISIJ Int. **50**(11), 1581–1589 (2010)
3. Vegman, E.F., Zherebin, B.N., Pokhvisnev, A.N., et al.: The Metallurgy of Iron: Textbook. IKTs Akademkniga, Moscow (2004)
4. Zhuravlev, V.F., Zhitomirskaya, V.I.: Binding properties of crystal hydrates of the sulfate type. J. Appl. Chem. USSR **23**, 115–119 (1950)
5. Harald, H.-K., Alexander, F., Stefan, H.: Briquetting of ferrous fines—saving resources, creating value. In: AISTech 2017 Conference Proceedings, vol. II, pp. 973–978
6. Bizhanov, A.M., Kurunov, I.F., Wakeel, A.Kh.: Behavior of extrusion briquettes (brex) in Midrex reactors. Part 2. Metallurgist (3), 112 (2016)
7. Berezhnoy, A.S.: Multicomponent Alkaline Oxide Systems, 200p. Kiev: Scientific Thought (1988)
8. Kurunov, I.F., Savchuk, N.A.: Status and prospects of non blast furnace iron metallurgy. Ferrous Metallurgy Information. p. 198 (2002)
9. Kikuchi, S., Ito, S., Kobayashi, I. et al.: KOBELCO Technology Review. No 29, pp. 77–84 (2010)

Chapter 6
Review of Alternative Applications of Stiff Extrusion Briquetting Technology

Another example of the industrial application of SVE technology for briquetting is the Coldry process, developed by the Australian Research Company Environment Clean Technology (ECT) for the agglomeration of the brown coal. At the first stage (the Coldry process), the brown coal is subjected to enrichment and drying, and then the brex are formed using a stiff vacuum extrusion process.

The results of theoretical and experimental research carried out at the organic chemistry faculty of the University of Melbourne in 1989 formed the basis of the Coldry-Matmor process [1]. The possibility of creating densified carbon was defined following an observation that was carried out at several Australian mines. It turned out that hardened bitumen-like road surfaces formed naturally soon after rain events when trucks had churned up brown coal fines with moisture when they entered and left the mine. In the days that followed a rain event, the road surfaces at the mine would harden like tarmac and no longer absorb water. A study into this phenomenon demonstrated that a process of low mechanical shear had occurred where brown coal mixed with a small amount of moisture and subject to low-level mechanical shear had substantially destroyed the coal porous structure and triggered a mild natural exothermic reaction process within the coal leading to the mobilization and subsequent evaporation of its moisture content. The low mechanical shear process fundamentally alters coal physical porous structure and to varying degrees the micro-chemical bonds within the coal, reducing moisture content between 10 and 14%, boosting calorific value over 5200 kcal/kg, and creating a new "densified coal" product that is hydrophobic, no longer prone to spontaneous combustion, readily transportable, and from a commercial and environmental point of view, a black coal equivalent [2]. The search for a briquetting technology that would be capable of achieving this effect led to the choice of stiff vacuum extrusion.

The overall process diagram for the Coldry process, one of the key components of which is SVE, is set out in Fig. 6.1.

This process starts with the screening and attritioning of the charge. A small amount of water is added to the attritioner, where the coal is sheared to form a coal paste. This intensive mixing initiates a natural chemical reaction within the coal

© Springer International Publishing AG 2018
I. Kurunov and A. Bizhanov, *Stiff Extrusion Briquetting in Metallurgy*,
Topics in Mining, Metallurgy and Materials Engineering,
https://doi.org/10.1007/978-3-319-72712-7_6

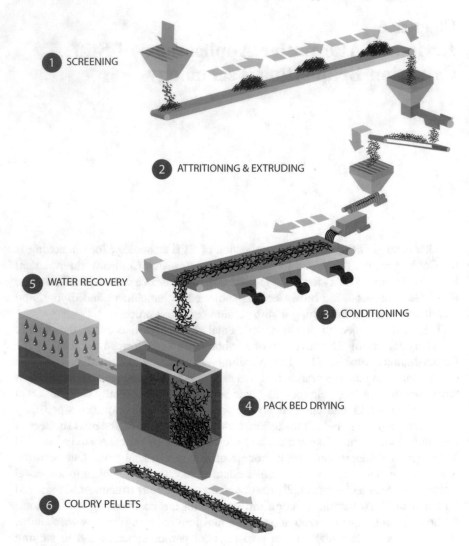

Fig. 6.1 The overall process diagram for the Coldry-Matmor process

which ejects both chemically trapped water and physically absorbed water within the coal pore structure. As already noted above, this process leads to a radical change in the structure of the brown coal when a shearing pressure is applied. A chemical analysis revealed the role of phenol residues in the formation of reaction centers within the structure of the modified coal. The activity of these residues is initiated by the attritioning movement, leading to new surfaces being revealed on the coal particles. The convergence of the reaction centers takes place during the course of the extrusion process, which means that at a later stage during the drying

Fig. 6.2 The appearance of SVE extruder (J.C. Steele & Sons, Inc), and the brex coming out of the die

of the brex new links can be formed. It is as a result of similar reactions that the densification of an agglomerated product occurs without a binder.

The plasticized mixture of coal- and iron-containing materials is fed into the extruder and comes out in the form of oblong brex (Fig. 6.2).

The newly formed brex are then placed in a plastic, corrugated conditioning belt (Fig. 6.3) in order to gain strength and to reduce the adhesiveness of the brex which is achieved by blowing warm air over the belt and the brex. The formation of new links within the coal components in the brex itself develops over the course of

Fig. 6.3 Brex on the conditioning belt

60 min. The consumption of air is determined using a calculation of the moisture content of the brex reaching 58% prior to their entry into the drying plant. The height to which the brex are allowed to pile up in the traverser is 15 cm.

Incoming moist brex from the conditioning belt are further dried to their ultimate moisture level within the packed bed dryer. Warm air from the heat exchangers removes the moisture rejected from within the coal pellets. The cross-linking reactions come to completion within the dryer, increasing the brex' strength to levels sufficient to withstand bulk transport. The effective duration of the drying process, according to ECT company information, is 48 h. The moisture content of the brex is between 10 and 12%.

Texas Industries, Inc. (TXI, USA) operated a number of lightweight aggregate (LWA) production plants. They used SVE technology to agglomerate ore fines (Fig. 6.4) for further reduction in their rotary kilns. The company operated two extruders at two LWA plants, one in Texas and one in Colorado. As of July 2, 2014, TXI became a wholly owned subsidiary of Martin Marietta Materials, Inc.

CE Minerals (Andersonville, Georgia, USA and Xiuwen, China): This company agglomerates high-grade bauxite through five of SVE extruders (Fig. 6.5). The extruded bauxite noodles or pellets are dried and fired in order to make high-grade alumina feedstocks which are used in making refractory products.

The producer of Bentonite Clay American Colloid (Mineral Technologies, MTX) uses J.C. Steele & Sons extruders to agglomerate and beneficiate dryer dust and low-grade Bentonite clay into brex of high-quality Bentonite through the use of additives and the shearing action of the extruder (Figs. 6.6 and 6.7). This company operates extruders at locations in Wyoming, South Dakota, China, and Thailand. The same equipment is being used also by WyoBen (Wyoming Bentonite).

Fig. 6.4 TXI lightweight aggregates produced by SVE

Fig. 6.5 High grade bauxite agglomeration by SVE

Fig. 6.6 Low-grade Bentonite clay agglomeration by SVE

Fig. 6.7 Low-grade Bentonite clay green brex transportation

Another application of this technology may be associated with the briquetting of red mud. The customer supplied a raw material for testing in J.C. Steele & Sons lab in Statesville (NC). The material consisted of moist, soft, sticky lumps ranging in size from fines to pieces as large as 50 mm (Fig. 6.8).

Fig. 6.8 Red mud sample material for SVE testing

Fig. 6.9 Brex made of red mud with 4% of Portland cement addition

The purpose of the test was to evaluate the suitability of the material for agglomeration by SVE. The test goals were to extrude the smallest practical pellet size and to use the minimum amount of binder (Portland cement) required. A test goal was to extrude pellets of the smallest practical diameter (Figs. 6.9 and 6.10). A pelletizing plate die for 6.4-mm-diameter brex was used. A 4% addition of

Fig. 6.10 Red mud brex after 7-day cure time

Portland cement was required to make the mix extrudable. With no water added, brex were well-formed but very soft with virtually no green strength.

After 7-day cure time, brex were crushed axially in order to determine axial compressive strength. The results of testing showed that this raw material, with the addition of Portland cement binder, is an excellent candidate for agglomeration by SVE.

Carolina Gypsum uses an extruder to agglomerate synthetic gypsum with a lignin binder to make fertilizer filler. Yearly production of this plant is about 20,000 tons per year (Fig. 6.11).

Universal Aggregates (PA, USA) agglomerates dry scrubber fly ash to make lightweight aggregate out of this pozzolanic material. The pellets are cured like concrete in a special curing vessel. Spray dryer ash, water, and other recycle materials are fed to a pug mixer where the materials are blended together. This mixing produces a uniformly blended loose, moist, granular material that feeds directly to an extruder. The extruder has an auger that subjects the material to further mixing and then forces the material through a die (metal plate with one or more drilled or specially shaped holes). Wet, "green" brex from the extruder are soft and must be transferred to a curing vessel for hardening (Fig. 6.12, [3]).

Fig. 6.11 Agglomeration of synthetic gypsum by SVE

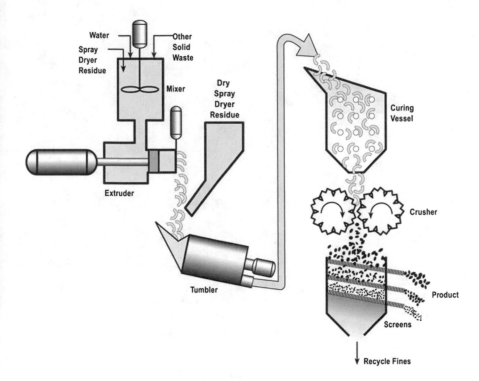

Fig. 6.12 Agglomeration of dry scrubber fly ash by SVE

References

1. Johns, B.A., Chaffee, A.L., Harvey, K.F.: The conversion of brown coal to a dense, dry, hard material. Department of Organic Chemistry, University of Melbourne, Victoria. In: Collaboration with A.S. Buchanan and G.A. Thiele at CRA Advanced Technical Development, Melbourne (1989)
2. Johns, R.B., Verheyen, T.V., Chafee, A.L.: Chemical characterization of Victorian brown coal lithotypes. In: Proceedings of International Conference on Coal Science. Dusseldorf 1981, pp. 863–868. Verlag Bluckauf (1981)
3. Ramme, B.W., Nechvatal, T., Naik, T.R., Kolbeck, H.J.: By-Product Lightweight Aggregates From Fly Ash. Report No. CBU-1995-11, November 1995, REP-352

Printed in the United States
By Bookmasters